Analytical Chemistry and Microchemistry

Analytical Chemistry and Microchemistry

Mass Spectrometry: Theory and Applications
William O. Nichols (Editor)
2021. ISBN: 978-1-53619-790-7 (Hardcover)
2021. ISBN: 978-1-53619-874-4 (eBook)

Aflatoxins: Biochemistry, Toxicology, Public Health, Policies and Modern Methods of Analysis
Spyridon Kintzios and Sofia Mavrikou (Editors)
2020. ISBN: 978-1-53616-785-6 (Hardcover)
2019. ISBN: 978-1-53616-786-3 (eBook)

Gas Chromatography: History, Methods and Applications
Percy Henrichon (Editor)
2020. ISBN: 978-1-53617-350-5 (Softcover)
2021. ISBN: 978-1-53617-352-9 (eBook)

A Textbook on Water Chemistry: Sampling, Data Analysis and Interpretation
A G S Reddy (Editor)
2020. ISBN: 978-1-53616-703-0 (Hardcover)
2020. ISBN: 978-1-53616-704-7 (eBook)

More information about this series can be found at https://novapublishers.com/product-category/series/analytical-chemistry-and-microchemistry/

Oscar M. Thygesen
Editor

Trace Metals

Sources, Applications and Environmental Implications

Additional color graphics may be available in the e-book version of this book.

Library of Congress Cataloging-in-Publication Data

ISBN: 978-1-68507-797-6

Published by Nova Science Publishers, Inc. † New York

Contents

Preface

Trace metals are necessary for the proper functioning of living organisms and are absorbed by the body through diet or environmental exposure. However, excessive intake of trace metals can cause health problems. As such, the study of the presence of trace metals in the environment and their effects on health is important. This volume includes four chapters that provide details about trace metals in various contexts. Chapter One explains the nutritional zinc requirements of humans and discusses the usefulness of different supplements in various applications. Chapter Two addresses the different aspects of metal-microbial interactions, focusing on soil and sediment ecosystems. Chapter Three addresses pollution of heavy metals, emission sources, health implications, and commonly used methods for assessment of pollution. Lastly, Chapter Four assesses possible changes in the geochemical behavior of chromium caused by sediment resuspension in a hypereutrophic estuary that receives domestic and industrial effluents daily.

Chapter 1 - Iron, zinc, copper, manganese, molybdenum, cobalt, and possibly, chromium are essential trace metals for humans. Zinc, in particular, is vital for many physiological processes that regulate endocrine and immune functions and control healthy brain development and activity. However, currently, it is estimated that 20 percent of the human population is at risk of zinc deficiency. Therefore, zinc supplementation became a significant focus in trace metal and health research, acting as a nutritional supplement to compensate for low zinc bioavailability and nutraceutical with pharmacological activity. This chapter will provide background to the nutritional zinc requirements of humans, introduce zinc supplements such as inorganic and organic supplements and discuss their differences and usefulness in various applications. In addition, the chapter will summarize scientifically proven beneficial effects of zinc supplementation and highlight new potential areas for zinc supplementation as a prevention and treatment strategy.

Chapter 2 - Modern globalisation has escalated the sources of heavy metal pollution in human-centered natural habitats. Heavy metals are persistent and are toxic to all forms of life. So, metal detoxification and restoration of metal-contaminated sites are very significant. Detoxification with microbes is very relevant as it is a cost-effective and natural method. Microorganisms living in already contaminated environments are often well adapted to survive in the presence of existing contamination. Microorganisms like algae, bacteria, and fungi can detoxify trace metals by bioremediation. Many reports on metal-microbe interactions highlight its important role in eradicating heavy metals from the ecosystem in an eco-friendly way through removal, detoxification, and recovery of organic and inorganic metals. There is a long quandary on how microbes link with metals in both natural and manmade environmental or biogeochemical cycling of metals by microorganisms. This chapter, therefore, attempts to address the different aspects of metal-microbial interactions, focusing on soil and sediment ecosystems.

Chapter 3 - With the rapid industrialization and economic development, heavy metals are continuing to be introduced to soils and sediments through fertilization, irrigation, rivers, runoff, atmospheric deposition and point sources. Additionally, activities such as metal mining, refining, and refinishing by products also contribute to the introduction of heavy metals in the environment. All these activities decrease the capability of the environment to support life thus threatening people, animal and plant health. Health implications associated with heavy metal exposure include those that disturbs nervous, blood forming, cardiovascular, renal and reproductive systems. Furthermore, accumulation of heavy metals in soil diminish quality of soil, cause crop yield decrease and affect the quality of agricultural products. It is important to appraise the concentration of heavy metals in the environment as they are toxic, persistent and non-degradable. Pollution indices are effective in appraisal of soil pollution with heavy metals, monitoring quality of soil and ensuring future sustainability. This chapter seeks to address the pollution of heavy metals, emission source, health implications and commonly used methods for assessment of pollution and for health risks assessment.

Chapter 4 - The resuspension of contaminated sediments in the water column has been recognized as an important process of metal pollutants remobilization in historically contaminated estuaries, which can change the concentration and bioavailability of these elements. The aim of this study was to assess possible changes on the geochemical behavior of chromium (Cr) caused by sediment resuspension in the area of a hypereutrophic estuary

(Guanabara Bay, Brazil) that receives domestic and industrial effluents daily during the recent decades. This study evaluated bioavailability change (BC) for Cr in estuarine sediments from Iguaçu River (located within the most impacted Guanabara Bay area), in response to laboratorial sediment resuspension experiments. The responses of sediments layers from different depth intervals were compared, since dredging activities usually promote resuspension of sediments removed from variable depths. Performed evaluations on the anthropogenic interference on sediment quality also included ecological risk index (E^if) estimates. Chromium concentrations obtained using a weak acid extraction (in a 1 mol L^{-1} HCl solution) were considered as the reactive (bioavailable) Cr phase. After resuspension along different time intervals, the uppermost sediment layers showed higher Cr concentrations in comparison with the non-resuspended control sediment. Some were above the Effect Range Low (ERL) sediment quality guideline, suggesting risks of adverse biological effects. These findings indicate increased potential bioavailability of the metal after resuspension. The E^if indicated low risk for Cr in all depth interval. The combined use of risk indices can be a useful tool for a more adequate management of dredging activities, helping in the prediction of contamination risks.

Chapter 1

Zinc Supplementation in Health and Disease

Sakshi Hans[1,2], MSc,
Janelle E. Stanton[1,2], MSc,
Eibhlís M. O'Connor[1,2,3,4], PhD and
Andreas M. Grabrucker[1,2,3,*], PhD

[1]Department of Biological Sciences, University of Limerick, Limerick, Ireland
[2]Bernal Institute, University of Limerick, Limerick, Ireland
[3]Health Research Institute (HRI), University of Limerick, Limerick, Ireland
[4]APC Microbiome Ireland, University College Cork, Cork, Ireland

Abstract

Iron, zinc, copper, manganese, molybdenum, cobalt, and possibly, chromium are essential trace metals for humans. Zinc, in particular, is vital for many physiological processes that regulate endocrine and immune functions and control healthy brain development and activity. However, currently, it is estimated that 20 percent of the human population is at risk of zinc deficiency. Therefore, zinc supplementation became a significant focus in trace metal and health research, acting as a nutritional supplement to compensate for low zinc bioavailability and nutraceutical with pharmacological activity. This chapter will provide background to the nutritional zinc requirements of humans, introduce zinc supplements such as inorganic and organic supplements and discuss their differences and usefulness in various applications. In addition, the

* Corresponding Author's E mail: andreas.grabrucker@ul.ie.

In: Trace Metals: Sources, Applications and Environmental Implications
Editor: Oscar M. Thygesen
ISBN: 978-1-68507-797-6

chapter will summarize scientifically proven beneficial effects of zinc supplementation and highlight new potential areas for zinc supplementation as a prevention and treatment strategy.

Keywords: zinc, trace metals, health, supplementation, nutrition, RDI, zinc deficiency

Introduction

The essential trace metal zinc is involved in various biological processes in living organisms. It is the second most abundant trace metal in the body after iron. It is also the most common divalent cation after calcium (Ca^{2+}) (Zhang et al. 2012). The average human body contains about 2-3 g of zinc (Zn^{2+}), with the highest amount of zinc present in tissues such as skeletal muscle and bone (Maret and Sandstead 2006; Hernandez-Camacho et al. 2020). Men tend to have higher zinc levels than women. A much smaller proportion of zinc is present in the serum, accounting for around 0.1% of the body's total zinc content. Normal plasma zinc concentrations are approximately 12-15 µmol/L (Hess et al. 2007).

In tissues, nearly all zinc ions (about 90%) are bound to proteins. However, free Zn^{2+} has a physiological role and can act as an intracellular signaling ion. In addition, zinc may also serve as an extracellular signaling ion, for example, after exocytosis of zinc-containing synaptic vesicles from neurons. These zinc-containing vesicles are mainly present in the glutamatergic neurons of the brain's cortex, hippocampus, and amygdala.

Zinc is involved in various physiological processes in organ systems such as the gastrointestinal, endocrine, immune, and nervous systems. This great variety of roles is mediated by the ability of zinc to bind over 300 enzymes and over 2500 transcription factors in the form of zinc finger proteins (ZnF) (King et al. 2015); (Bagherani and R Smoller 2016). The physiological roles of protein-bound Zn^{2+} can be structural, catalytic, and regulatory. For example, zinc is a catalytic factor for over a hundred enzymes that control growth, proliferation, DNA synthesis, and other functions. In addition, it is involved in the mechanisms of apoptotic pathways and also influences the regulation of the growth factor IGF-1, a hormone that promotes bodily growth (MacDonald 2000). Zn^{2+} can also have an antioxidant role by inhibiting the production of reactive oxygen species (ROS) (Lee 2018).

It is therefore not surprising that zinc deficiency is associated with various health problems. For example, evidence from animal models has shown that zinc deficiency causes depression-like symptoms, and studies found that zinc supplementation may have benefits such as enhancing the effectiveness of antidepressants (Szewczyk et al. 2002). However, the precise mechanisms of action have not been fully explored yet. In human patients, low zinc status is also linked to depression (Gronli et al. 2013). Clinical studies have uncovered an inverse relationship between the severity of depression symptoms and plasma zinc levels (Swardfager et al. 2013). A randomized, controlled clinical study found that zinc supplementation significantly reduced symptoms of depression in depressed patients (Lai et al. 2012). In humans, low zinc levels are also associated with anorexia, loss of smell and taste senses, immune dysfunction (Chasapis et al. 2012), and mental health issues such as Autism Spectrum Disorders (ASD) and Attention Deficit Hyperactivity Disorder (ADHD). Reduced plasma zinc levels have been detected in the blood of children with ASD and ADHD and other tissues (hair, nails, and urine). Maternal and fetal zinc deficiency have also been connected with the psychiatric disorder schizophrenia.

In addition, zinc therapy may be helpful in the treatment of anxiety disorders, possibly by exerting an antioxidant effect (Russo 2011). Zinc supplementation is therefore being explored as a potential therapeutic option to treat psychiatric disorders, especially as it has been shown that Zn^{2+} notably influences synaptic plasticity, which describes the structural and biochemical modifications occurring in neuronal synapses over time (Grabrucker et al. 2011) that are key to healthy brain development and functioning.

In the immune system, Zn^{2+} supports T-cell division, primary and secondary antibody response, and production of Cu, Zn, superoxide dismutase (SOD), which is involved in the antioxidant defense system. In line with this, zinc levels are shown to be linked with disease progression and mortality rates in human immunodeficiency virus (HIV), although the exact significance of zinc levels and HIV is not fully clear yet. A study showed that excessive zinc intake is associated with a faster progression of the disease and a higher risk of death (Baum et al. 2000). At the same time, low plasma levels of zinc are linked with a reduced survival rate. (Baum et al. 2003). The anti-viral activity of zinc has also been confirmed for a range of viral infections, such as the common cold and herpes simplex virus (Read et al. 2019). A study showed that zinc salts could efficiently inhibit clinical isolates of the herpes simplex virus (HSV) (Arens and Travis 2000). In addition, the World Health Organization (WHO, 2020) reports zinc deficiency to be responsible for 16%

of all deep respiratory infections. This has led to several studies investigating zinc as a possible therapeutic agent against SARS-CoV-2 and other respiratory diseases (Read et al. 2019; Wessels et al. 2020). Zinc seems to have a protective effect against viruses through multiple mechanisms. These include inhibiting viral replication and preventing entry of the virus by preserving the integrity of the respiratory epithelium. For example, the interaction of the SARS-CoV-2 spike protein with enzymes on the host cell surface is blocked by zinc, which inhibits the expression of ACE2 (angiotensin-converting enzyme 2) (Devaux et al. 2020).

Zinc can also restore balance to the immune system in response to the destruction caused by inflammation in the lung epithelia (Bao and Knoell 2006). Furthermore, patients with diseases like asthma (Chen et al. 2020), chronic obstructive pulmonary disease (COPD) (Kirkil et al. 2008), autoimmune disorders (Sanna et al. 2018), etc., have low plasma zinc levels (i.e., there is an overlap in risk groups for severe Covid-19 and zinc deficiency), pointing at an association between zinc deficiency and susceptibility to virus infection (Wessels et al. 2020).

Given these diverse roles of zinc in human health and development, zinc status is tightly controlled in tissues, and sufficient zinc supply is necessary throughout the whole life.

Nutritional Requirement of Zinc in Humans and Dietary Sources of Zinc

Adequate zinc levels are required in all stages of life, such as growth and development during pregnancy and childhood, adulthood, and healthy aging. Plasma/serum zinc concentration and other biomarkers of zinc adequacy, deficiency, and excess are not useful for estimating dietary reference values (DRVs) for zinc which requires a factorial approach involving two stages. The first stage involves an estimation of physiological requirements, defined as the minimum quantity of absorbed zinc needed to match losses of endogenous zinc and to meet any additional requirements for absorbed zinc, e.g., for growth in healthy, well-nourished infants and children and in pregnancy and lactation. The second stage determines the quantity of dietary zinc available for absorption that is needed to meet these physiological requirements ('Dietary Reference Values for nutrients Summary report' 2017; Ranasinghe et al. 2018). In cases of zinc deficiency, dietary zinc supplementation may support healthy pregnancy and reduce the risk of preterm births (Ota et al.

2015). However, daily zinc requirement differs among population groups, with pregnant and lactating women having a notably higher recommended intake than men and non-pregnant or lactating women ('Dietary Reference Values for nutrients Summary report' 2017). This is owing to the increased demand for zinc for fetal development and breastmilk production. The Population Reference Intake (PRI) for adult women ranges from 7.5 to 12.7 mg zinc per day. It is 9.4 to 16.3 mg/day for men, with an additional intake of 1.6 and 2.9 mg/day recommended for pregnant or lactating women, respectively (EFSA Panel on Dietetic Products, 2017). Young children and infants are more vulnerable to zinc deficiency due to the increased zinc requirement during their growth and development (Table 1).

Table 1. Daily zinc requirement (mg/day) across genders and age groups, adjusted for different levels of phytate intake (LPIs)

Population	Level of phytate intake (LPI) (mg/day)	Zinc requirement (mg/day)
Females 7-11 mo.	-	2.9
7-10 yr		7.4
15-17 yr		11.9
Adult females (18-24 yr)	300	7.5
>25 years	600	9.3
	900	11.0
	1,200	12.7
Males 7-11 mo.	-	2.9
7-10 yr		7.4
15-17 yr		14.2
Adult males (18-24 yr)	300	9.4
>25 years	600	11.7
	900	14.0
Pregnancy	+1.6	
Lactation	+2.9	

Adapted from: Dietary Reference Values for nutrients Summary report', (2017) *EFSA Supporting Publications*, 14(12), available: https://dx.doi.org/10.2903/sp.efsa.2017.e15121.

*We have not included adjustments for phytate intake since the Zn absorption for children was based on data from mixed diets containing variable quantities of phytate.

The risk of zinc deficiency is not evenly distributed globally, with people living in low-income or developing countries at higher risk of zinc deficiency (Black 2003). The global risk of dietary zinc deficiency is estimated to be about 17%, with regions in South Asia, sub-Saharan Africa, and Central America at the highest risk of zinc deficiency (Wessell et al. 2012). For example, the prevalence of zinc deficiency is about 30% in South Asia, while in contrast, it is estimated at 7.5% in industrialized countries (Western European nations and North America).

This disparity could be due to many factors, such as lower meat consumption and the prevalence of vegetarian diets in severely affected countries. Certain foods are particularly rich sources of zinc, for example, red meat, fish, eggs, legumes, grains, and grain-based products. However, it is shown that zinc absorption is 15-26% from vegetarian and lacto-ovo vegetarian diets, and thus much lower than the 33-35% from omnivorous mixed diets (Gibson 1994). Indeed, the growing popularity of vegan, plant-based diets could prove problematic in the future in terms of disease risk associated with sub-optimal zinc intake. In addition to the zinc content of foods, other factors govern zinc levels in the body, such as the bioavailability of dietary zinc that is impacted by the presence of inhibitors of zinc absorption, such as copper, phytates, and folic acid (Gibson et al. 2018). Phytate is a known inhibitor of zinc absorption and is often abundant in plant-based diets. Thus, the ratio of phytic acid to zinc (Phy:Zn ratio) determines the bioavailability of dietary zinc. Staple foods such as unrefined maize flour, brown rice, and legumes have high Phy:Zn ratios (Gibson et al. 2018). Studies also show that zinc and folic acid undergo an inhibitory interaction in the intestine (Ghishan et al. 1986; Gibson et al. 2018).

Other factors have been identified that cause zinc deficiency. Multiple studies have shown that consumption of copper and iron supplements affects zinc uptake in the intestine (Gibson 1994; O'Brien et al. 2000). Specific methods of processing plant foods such as cereals can modify their phytate content, which has implications for the bioavailability of zinc (Gibson et al. 2018). Thus, the bioavailability of zinc is influenced by several factors, such as the dietary source of this micronutrient. However, zinc deficiency worldwide is high, with zinc deficiency a growing cause for concern in developing countries. Especially pregnant and lactating women are at high risk of developing low zinc status. Therefore, zinc supplementation is getting more and more in focus for improving and maintaining health and wellbeing (Fourie et al. 2018; Corona et al. 2010; Maret and Sandstead 2006).

Zinc Absorption

The absorption of zinc is primarily through the brush border cells of the small intestine. Entry of zinc occurs via the apical side of the epithelial cells of the gastrointestinal tract (enterocytes). Zinc absorption by the enterocyte is regulated in response to the quantity of bioavailable zinc ingested. To facilitate this, several transport systems exist to deliver zinc ions across biological membranes, and these zinc transporter proteins have an essential role in controlling zinc homeostasis. Albumin is the major transporter of zinc in both portal and systemic circulation. The quantity of zinc secreted into and excreted from the intestinal tract depends on body zinc concentrations. Accordingly, the enterocytes express various zinc transporter genes. The two major groups of zinc transporters are the 14 zinc-importers ((ZIP) Zrt, Irt-like proteins (*Slc39a*)) which deliver zinc across the plasma membrane into the cytosol, and the zinc transporter family of 10 genes (ZnT (*Slc30a*)) that export zinc out of the cytosol (Chasapis et al. 2012). In addition, another group of proteins, the metallothioneins, can bind free zinc present in the cytoplasm and regulate its levels (Maares and Haase 2020).

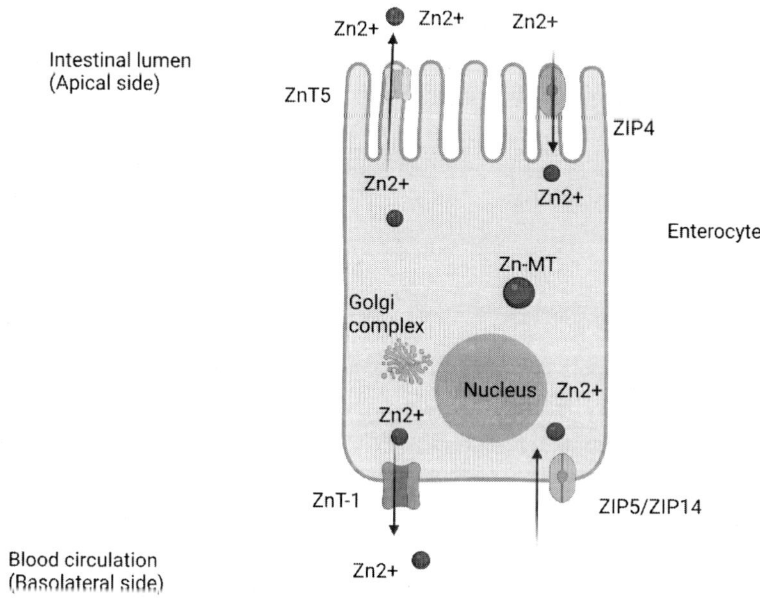

Figure 1. Overview of zinc absorption in the small intestine. Import and export of Zn^{2+} via transmembrane receptors located in the enterocytes are depicted. Transport of extracellular Zn on the apical side of the enterocytes is mediated by the ZIP4 protein family, while ZnT proteins carry Zn out of the cytosol. Much of the cellular Zn is complexed to metallothionein (MT), which also regulates the intracellular levels of free Zn.

A critical zinc transport protein in the process of zinc absorption is ZIP4 (SLC39A4). The zinc absorption disorder *Acrodermatitis enteropathica* (AE) results from mutations in the ZIP4 gene. Export of cellular zinc is mainly mediated by ZnT1, which is localized at the basolateral membrane of enterocytes (Cousins 2010).

Zinc Supplements

The efficacy of zinc supplementation depends not only on the amount of supplement taken but also on its bioavailability. Bioavailability is a measure of the maximum rate to which a dietary zinc supplement can provide the body with physiologically utilizable zinc. It is calculated by the maximum zinc absorption from the GI tract into the blood circulation and the maximum possible rate of metabolic utilization. Metabolic utilization refers to how much absorbed zinc is retained and what proportion of absorbed zinc is excreted. In general, zinc bioavailability depends primarily on absorption (Schlegel and Windisch, 2006), which is influenced by several factors, among them the type of zinc formulation.

In general, zinc formulations are differentiated into inorganic zinc and organic zinc supplements (Table 2). Inorganic zinc, such as zinc salts, dissociate into "free" zinc ions in solution (e.g., in the GI system, i.e., $ZnCl_2 \rightarrow Zn^{2+} + 2Cl^-$). In contrast, organic zinc such as zinc amino acid chelates stabilize the zinc ion and will not immediately release "free" zinc. Comparative studies on the absorption of different dietary zinc formulations, such as zinc chloride, zinc sulfate, zinc fumarate, and zinc histidine, revealed high rates of around 95% absorption (Weigand and Kirchgessner 1979). However, these values were measured under the condition of low zinc status. Under regular zinc supply, organic zinc supplements seem to be more bioavailable (Ren et al. 2020; Suo et al. 2015; Schlegel and Windisch 2006). The variations in bioavailability between inorganic and organic zinc sources are likely caused by differences in how the formulations interact with other dietary constituents such as zinc uptake inhibitors (i.e., phytic acid) and the mode of uptake (transport mechanisms).

For example, organic zinc formulations, such as zinc proteinate and zinc amino acid chelates/conjugates, have been shown to have higher relative bioavailability than that of inorganic zinc supplements, such as zinc oxide and zinc sulfate (Wedekind and Baker 1990; Wedekind et al. 1992; Du et al. 1996; Cao et al. 2000). Zinc integrated into the ring structure in zinc amino acid conjugates (ZnAAs) is protected from interactions with uptake inhibitors such as phytic acid in the GI tract (Wedekind et al. 1992).

Table 2. Dietary zinc supplements and their classification

Zinc formulation	Examples
Inorganic zinc supplements	Zinc sulfate ($ZnSO_4$), Zinc oxide (ZnO), Zinc chloride ($ZnCl_2$), Zinc gluconate ($C_{12}H_{22}O_{14}Zn$), Zinc acetate ($ZnC_4H_6O_4$), Zinc picolinate ($C_{12}H_8N_2O_4Zn$), Zinc citrate ($C_{12}H_{10}O_{14}Zn_3$), Zinc orotate ($C_{10}H_6N_4O_8Zn$)
Organic zinc supplements	Zinc-polysaccharide, Zinc amino acid conjugates (ZnAAs, e.g., ZnMet, ZnGlu, ZnLys), Zinc chelate of ethylenediamine (Zinc-EDA-Cl)
Biologically organic zinc supplements	Zinc-enriched yeast, seaweed zinc extracts, etc.

ZnAAs like Zn-Methionine (ZnMet) coordinate zinc with covalent bonds, and the chelate is stable in the small intestine. This may minimize the formation of a Zn-phytate complex and allow more zinc to be absorbed (Behjatian Esfahani et al. 2021). In contrast, inorganic zinc formulations such as $ZnSO_4$ readily dissociate in the stomach and intestine, allowing zinc ions to form Zn-phytate complexes, thereby lowering the absorption of zinc (Ren et al. 2020).

Furthermore, ZnAAs are absorbed using amino acid transporters instead of the classic zinc importers/exporters such as SLC30A and SLC39A family members (Sauer et al. 2017). Being absorbed by different routes also lowers competition with other divalent metals for zinc transporters and zinc-buffering proteins such as MTs. Several studies confirm these advantages of organic zinc supplements. For example, studies have shown that the zinc retention and bioavailability of zinc glycinate is 15-30% higher compared to zinc sulfate (Schlegel and Windisch, 2006). In line with this, dietary supplementation of zinc as an organic or inorganic form both showed increased growth performance in chicken. However, organic zinc supplementation resulted in higher tissue zinc accumulation than inorganic zinc supplementation (Ao et al. 2009).

In pig feedstuffs, in general, high levels of inorganic zinc supplements ($ZnSO_4$, ZnO) are used (Buff et al. 2005) but these can negatively modify gut microbiota composition and may also result in significant excretion of zinc into the environment (Liu et al. 2020). In contrast, organic zinc formulations can be given at lower doses due to their higher bioavailability (Zhang et al.

2018a). For example, Xie et al. (2019) reported that piglets fed a diet supplemented with a lower amount of organic zinc (ZnMet) replacing 100 mg/kg $ZnSO_4$ had no adverse effects on health and performance indicators. Besides, lower dietary concentrations of an organic zinc formulation (300 or 450 ppm zinc as zinc-polysaccharide) were found to maintain growth performance compared with pharmacological concentrations of inorganic zinc (2,000 ppm zinc as ZnO) (Case and Carlson, 2002; Buff et al. 2005). Furthermore, Li et al. (2019) found that in laying hens, dietary ZnMet had a more beneficial effect on zinc accumulation in tissues, intestinal morphology, and metallothionein (MT) gene expression in the intestinal tract compared to $ZnSO_4$ supplementation.

Another type of organic zinc supplement, termed biologically organic zinc, such as zinc-enriched yeast (ZnY), is becoming more commonly used. In biologically organic zinc, zinc is also protected from forming insoluble complexes in the GI tract. For example, ZnY is a complex of proteins, peptides, and amino acids produced by yeast fermentation (Zhang et al. 2018b). In a study that compared zinc sulfate, zinc glycinate, and ZnY at a low dose, the different formulations showed equivalent bioavailability based on plasma and tissue zinc levels. However, ZnY was better retained than zinc sulfate and zinc glycinate based on zinc excretion levels (Zhang et al. 2018b).

To maximize zinc absorption, a combination of inorganic and organic zinc formulations seems highly beneficial. For example, supplementing 50 mg $ZnSO_4$ plus 25 mg ZnMet to piglets exerted the beneficial effects of zinc (Xie et al. 2019). This effect may be explained by the different modes of uptake and absorption of inorganic and organic zinc. In addition, it was shown that organic zinc supplements such as ZnAAs are taken up by amino acid transporters (Sauer et al. 2017). Thus, while inorganic zinc may saturate zinc uptake transporters in the GI tract, such as ZIP4 (SLC39A4), additional zinc may be absorbed at the same time through amino acid transporters in the form of organic ZnAAs. Thus, using a combination of inorganic and organic zinc supplements may allow reaching higher tissue zinc levels which may be needed for pharmacological effects of zinc.

This difference in uptake should also be considered when using zinc as a nutraceutical. For example, individuals with *Acrodermatitis enteropathica* (AE) frequently have mutations in ZIP4. Therefore, high zinc levels are needed to balance the effects of non-functional ZIP4 by exploiting other zinc importers present in the GI tract, such as ZIP1 and ZIP2. However, it was shown that uptake of ZnAAs was not affected by mutations in ZIP4, as ZnAA uptake is predominantly mediated by amino acid transporters (Sauer et al.

2017). Thus, organic zinc supplements such as ZnAAs may be more suitable for individuals with AE. Similarly, it has been shown that individuals with Phelan McDermid Syndrome (22q13.3. deletion syndrome), a developmental disorder often associated with autistic behaviors, have an impaired ZIP transporter system in the GI tract and frequently present with zinc deficiency (Pfaender et al. 2017; Grabrucker et al. 2014). Thus, also here, ZnAAs may be a more suitable zinc formulation. In addition, inflammatory processes in the GI tract may affect the zinc transporter system differently than amino acid transporters.

Thus, taken together, future applications will need to consider the required tissue concentration and intra- vs. extracellular effects, guiding the selection of the zinc supplement or combinations of supplements, and explore tissue-specific targeted delivery.

Zinc as a Nutraceutical

Human disease is influenced by many factors, both genetic and environmental. The impact of minerals and vitamins on health and development has been vastly studied in recent years. A particular emphasis has been put on zinc and its role in various conditions and diseases such as depression, diabetes mellitus, diarrhea, skin problems, and viral infection such as those causing the common cold and Covid-19 (Chukwuma et al. 2020; Costagliola et al. 2021; Swardfager et al. 2013; Wilson et al. 2006). As zinc supplementation has been shown to decrease oxidative stress and regulate inflammatory cytokines in disease states, zinc supplementation likely affects the severity and progression of many more diseases (Bao et al. 2008; Kelishadi et al. 2010; Guo and Wang, 2012; Banupriya et al. 2020).

Diabetes is a growing problem in society globally, with studies investigating both prevention and management of the disease. A particular focus has been put on the effects of zinc supplementation on blood glucose control which may be particularly beneficial for patients with type-2 diabetes (Islam et al. 2016). In fact, zinc supplementation in individuals with type-2 diabetes (ranging from 40 mg-200 mg/day) was reported to have a favorable effect on total cholesterol and triglyceride levels (Ashaghi et al. 2020). Furthermore, zinc is thought to play a principal role in the transport and formation of insulin granules in type 1 diabetes, also referred to as autoimmune diabetes. Hence, the addition of zinc supplements has been shown to reduce the risk of inflammation and promote insulin production (Sanna et al. 2018).

Diabetes mellitus management depends on several factors, including glycemic control. The level of glycated hemoglobin (Hba1c) indicates an individual's average blood sugar levels. A meta-analysis of studies investigating zinc supplementation in diabetes reported lower Hba1c levels by nearly 0.6% in patients with diabetes mellitus. It also reduced total cholesterol, LDL-C, and HDL-c levels in patients, indicating that it can also affect lipid metabolism (Jayawardena et al. 2012). Another investigation of a randomized, double-blinded Phase 2 clinical trial found similar results whereby pre-diabetic adults who received a daily dose of 20 mg of zinc showed reduced glucose levels and insulin resistance compared to a placebo cohort. A significant reduction was noted for TC and LDL-c levels as well. The latter is a predictor of developing cardiovascular disease and of the rate of disease progression (Ranasinghe et al. 2018). Intriguingly, an islet-restricted zinc transporter known as ZnT8 (SLC30A8) has recently been linked to the regulation of insulin secretion, which may influence the development of the disease (Yoshikawa et al. 2001; Rutter, 2010; Jayawardena et al. 2012).

The use of oral zinc supplementation in cases of viral-induced diarrhea, especially in children older than six months, has been supported by current literature (Afolabi et al. 2019; Bajait and Thawani, 2011; Krebs, 2013; Lukacik et al. 2008; Trivedi et al. 2009). Zinc supplementation is beneficial in treating and preventing diarrhea among children with an 18% decline of diarrhea symptoms and 14% decreased risk in the incidence of diarrheal episodes (Aggarwal et al. 2007; Lukacik et al. 2008). During this developmental period, children require zinc for efficient neurodevelopment (Adamo and Oteiza, 2010; Colombo et al. 2014), and zinc plays a role in influencing the gut microbiome (Lazzerini, 2016; Yu et al. 2021).

The role of zinc supplementation in treating skin diseases has also been investigated. The integrity of the skin through the barrier and immune mechanisms relies on many factors, including zinc bioavailability. The results of low serum, hair, and erythrocyte zinc levels can lead to conditions such as acrodermatitis, acne vulgaris, vitiligo, and psoriasis. Adding zinc to topical treatments and zinc supplements may improve or control these conditions (Zeng et al. 2014; Tavakoli et al. 2018; Gray et al. 2019; Lei et al. 2019). For example, in acne treatments, the use of zinc-containing topical treatments showed significant improvements compared to individuals who received placebo treatments (Yee et al. 2020).

Topical remedies containing zinc, such as zinc oxide, have historically been used to treat eczema, rosacea, psoriasis, acne vulgaris, and minor skin irritations. Salts of zinc such as zinc sulfate, zinc gluconate, and zinc acetate appear to be effective in managing inflammatory acne in combination with antibiotics such as erythromycin or alone (Gupta et al. 2014). However, the results of these studies are mixed, and the mechanism by which zinc exerts this anti-acne effect is not fully clear. An explanation could be the anti-inflammatory properties or inhibition of lipases produced by Propionibacterium acnes (Bae et al. 2010). In addition, zinc used alone or in combination with nicotinamide is a newly emerging treatment with the benefit of bypassing traditional antibiotic use, which may be ineffective due to antibiotic resistance in P. acnes.

Acrodermatitis enteropathica is a rare autosomal recessive disorder caused by mutations in the zinc transporter gene *Zip4*. This mutation leads to impaired zinc absorption in the small intestine (in the duodenum and jejunum). Therefore, zinc replacement therapy is required to manage this condition, with doses typically at 1-3 mg/kg of elemental zinc each day (Kilic et al. 2012) (Barnes and Moynahan 1973)

Zn administration has also been associated with shortened respiratory infections when taken within the first 24 hours of symptoms. Intake of zinc supplements in the form of gluconate lozenges saw a reduction in the duration of the common cold. 86% of subjects who received a 23 mg zinc tablet were asymptomatic compared to the placebo group (46% asymptomatic) after seven days of treatment (Eby et al. 1984). Variations in results over the years may be due to the levels of free Zn ions in lozenges and supplements, and the localized effects zinc lozenges may have, which may be varied by the administration/ingestion of the lozenges. Additionally, the concentrations of zinc in the examined tablets differed from 20 mg-200 mg across studies, including participants with self-reported cold incidences and clinical pneumonia cases (Hemilä, 2017; Wang et al. 2020). Various trials have been analyzed in a meta-analysis concluding a 34% decrease in length of time of nasal discharge, along with a 46% reduction of time in which a cough is exhibited in patients receiving high zinc doses of 80-92 mg/day (Hemilä and Chalker, 2015).

Additionally, symptoms of virus-causing illnesses such as Covid-19, which is caused by SARS-CoV-2, have also been shown to diminish more rapidly with the addition of zinc supplements, along with reduced pneumonia morbidity of 19% in developing countries (Yakoob et al. 2011; Wang et al. 2020; Wessels et al. 2020). According to a recent study, the prevalence of zinc deficiency is positively correlated with the number of Covid-19 cases per million in Asian countries, although no significant correlation with the number of deaths per million was found. However, this study also found an inverse relationship between zinc deficiency and the number of both COVID-19 cases or deaths due to covid per million in European populations (Ali et al. 2021). Although, this may be explained by the lower rate of zinc deficiency in Europe compared to Asia, as well as the high number of COVID-19 cases.

Treatment plans of patients with Covid-19 who present with underlying long-term conditions which add to the risk of more serious Covid-19 symptoms, including but not limited to cardiovascular disease, chronic obstructive pulmonary disease, and asthma, require further interventions and preventions to protect the vulnerable (Karim et al. 2021; Song et al. 2021). The effects zinc deficiency has on immune function, such as reduced lymphocyte numbers, decreased antibody production, and diminished activity of immune cell functions such as phagocytosis, adds to the epidemiology in patients with more severe Covid-19 symptoms (Wessels et al. 2017; Jothimani et al. 2020; Mossink, 2020). Thus, the use of zinc supplements such as zinc acetate along with antivirals such as nitazoxanide, ribavirin, and ivermectin in cases of mild Covid-19 have shown improved times of associated symptoms (Elalfy et al. 2021; Finzi and Harrington, 2021).

Age-related macular degeneration (AMD) is a common disease amongst persons aged 50 to 60 years and over, causing everyday activities such as reading to become increasingly difficult. There are various causes of the diseases, such as genetic and non-genetic factors. A regime of antioxidants and high-dose zinc supplements are usually recommended to prevent and improve symptoms of neovascular AMD (Vavvas et al. 2018). Zinc supplementation taken along with vitamin C, D, and B12 has decreased the risk of AMD and improved macular function in patients. The intake of carotenoids and antioxidants such as zinc has shown a slower advancement and improved visual perception (Beatty et al. 2013; Zampatti et al. 2014; (Vishwanathan et al. 2013).

The US National Eye Institute (NEI) endorses using the AREDS2 nutrient formulation for preventing progression from intermediate to advanced AMD (Chew 2013; Chew et al. 2014). Zinc is highly concentrated in the outer retinal and particularly in the pigmented retinal pigment epithelium (RPE). However, zinc levels decrease with aging and are particularly low in those with AMD. AMD occurs as RPE pigmentation and zinc levels decrease, but the relationship among these events is unknown.

Overall, zinc supplementation has a vast range of uses as a nutraceutical (Table 3), particularly in viral prevention and management. Additionally, its role as an antioxidant provides applications in numerous diseases and skin conditions. The presence of zinc deficiency in patients with various diseases tends to result in more severe symptoms. Thus, zinc supplementation individually and along with other anti-viral drugs, co-antioxidants and vitamins leads to improved health and immune defense.

Future Perspectives of Zinc as a Nutraceutical

There is mounting evidence that zinc deficiency is linked with neuronal dysfunction and is implicated in multiple disorders of the CNS (Gower-Winter and Levenson 2012; Szewczyk 2013). The importance of zinc homeostasis begins from the earliest steps of brain formation, i.e., neurogenesis or development of the neural tube. For example, it has been demonstrated that zinc deficiency affects CNS functioning, specifically in learning and behavioral areas, and is notably linked to ASD (Bhatnagar and Taneja 2001; Stanton et al. 2021).

ASD are a group of developmental disorders characterized by impairments in social functioning and speech, and the presence of repetitive behaviors. Although ASD has a strong genetic component, many environmental factors are suspected to increase the risk for developing it, notably prenatal zinc deficiency and maternal immune dysfunction. Zinc levels directly affect immune function, which is why lack of adequate Zn is a risk factor for developing ASD (Grabrucker 2012).

The role of zinc in maintaining the gut microbiome has been established, and zinc deficiency would consequently disturb the balance of the microbiota (Vela et al. 2015). Many individuals with ASD have gastrointestinal comorbidities or dysfunctions, such as diarrhea, constipation, and gastrointestinal reflux (Chaidez et al. 2014). This hints towards a gut-brain signaling pathway in ASD. Maternal zinc deficiency has been associated with depression-like behaviors and appears to be a risk factor for developing ASD (Grabrucker 2012; Grabrucker et al. 2014; 2016; 2017; Vela et al. 2015; Schoen et al. 2019; Daini et al. 2021). Maternal zinc deficiency could be owing to antagonistic interaction of other supplements with zinc, high consumption of inhibitors such as phytates, or intake of drugs that interfere with zinc absorption (Vela et al. 2015). Because of this, zinc supplementation in women of childbearing age could prevent the occurrence of ASD in their children.

Similarly, supplementation in young ASD patients could reduce the burden associated with comorbidities such as GI issues and immune dysfunction (Hagmeyer et al. 2018). As reported in a survey of the effectiveness of various nutraceuticals in the context of ASD, improvement in the severity of ASD symptoms was observed when zinc supplements were administered, with minimal side effects (Adams et al. 2021). Zinc therapy has been shown to attenuate oxidative stress and provide a neuroprotective effect against environmental insults. To illustrate, mouse models exhibited decreased lipid peroxidation in the cerebellum four months following supplementation with zinc (Bhalla et al. 2007). In line with this, ASD-like behavioral deficits in several mouse models for ASD were rescued by zinc supplementation. For example, ASD mouse models treated with valproic acid, Shank3-knockout mice, and maternal immune activated (MIA) mice all showed improvement in ASD-associated behaviors after Zn supplementation (Fourie et al. 2018; Vyas et al. 2020; Lee et al. 2015; Kirsten et al. 2015; Cezar et al. 2018).

Epilepsy and ADHD are comorbidities occurring relatively frequently with ASD (Besag 2018). The role of zinc in regulating epileptic attacks and seizures is controversial, with studies yielding conflicting results. However, a trial involving children with intractable epilepsy (IE) found that oral zinc supplements significantly brought down seizure rates in 31% of children (Saad et al. 2015).

The management of ADHD is reliant on dopamine-regulating prescription drugs. However, studies have shown a significant increase in effectiveness when the prescribed drugs are taken in parallel with zinc supplements (Kirby et al. 2001). Zinc has a crucial influence on melatonin production, which in turn regulates dopamine function. Thus, the prescription of ADHD regulating drugs and zinc supplements in individuals with zinc deficiency show increased improvements in ADHD etiology (Akhondzadeh et al. 2004). Epilepsy and ADHD were associated with low zinc status in individuals with Phelan McDermid Syndrome, an ASD (Grabrucker et al. 2014; Besag, 2018). However, larger-scale studies are required to establish whether zinc supplements are beneficial to treat epilepsy in combination with antiepileptic drugs or ADHD in ASD.

Overall, human studies examining zinc as a supplement for ASD patients have yielded mixed results. An effective dosage level of zinc from a suitable source needs to be identified in this respect. The use of inorganic zinc is controversial, as uptake of ionic zinc by Zip4 and Zip2 transporter proteins may be compromised in some ASD. This issue may be avoided by using organic forms of supplements, such as zinc-amino acid complexes (Hagmeyer et al. 2018). Transport of such complexes would be mediated by amino acid transporters in the enterocytes and result in a highly efficient rate of zinc uptake. More importantly, the timepoint of zinc supplementation is critical. Low zinc status was associated with ASD mostly early in life and may primarily impact during brain development *in utero*. Therefore, zinc supplementation very early in life, which will be dependent on better early detection of ASD, or maternal supplementation during pregnancy, may have much more potent effects than supplementation of older individuals with ASD.

Dysregulation in zinc homeostasis is also implicated in other neuropsychiatric disorders such as schizophrenia, depression, and ADHD (Chhabra et al. 2015) (Petrilli et al. 2017). In addition, recent research supports the role of zinc therapy in the treatment of a bipolar disorder. A study involving patients diagnosed with this disorder assessed their serum hepatocyte growth factor (HGF) levels prior to and after zinc therapy. The protein HGF is a regulator of cell growth and morphogenesis and is particularly significant in the development of the CNS. It is also a modulator of the neurotransmitter GABA, reduced levels of which are associated with bipolar disorder. The study results showed a significant reduction in symptoms post-therapy, and serum HGF returned to normal levels (Russo 2010). This may be due to reducing oxidative stress and damage to liver cells attributed to zinc therapy that upregulates cell proliferation genes, including HGF.

In line with this, another study demonstrated that neuronal cell proliferation is increased by zinc supplementation in rats and also prevents symptoms associated with traumatic brain injury (TBI) (Cope et al. 2016), notably depression. The typical depression-associated behavior anhedonia was studied using the saccharin test in rats with induced TBI. It was found that zinc supplementation improved cell proliferation in the hippocampus after TBI and reduced depression-like behaviors in rat models. Zinc also seemed to promote neuronal differentiation in the hippocampus. These findings represent a potential new therapy for TBI-induced depression, especially considering that depression is the most common consequence after TBI and seems resistant to conventional antidepressants (Silverberg and Panenka 2019).

The risk of zinc deficiency may increase in advanced age due to worsened zinc absorption and lifestyle factors, such as a diet containing less bioavailable zinc (Sauer et al. 2020). Because of this, zinc supplementation could be beneficial in the elderly population and combat the risk of developing age-related diseases. For example, it has also been suggested that an imbalance of trace metal homeostasis, notably of zinc and copper, plays a role in the pathology of neurodegenerative diseases, such as Alzheimer's disease (AD). Zinc supplementation has been demonstrated to have beneficial effects such as preventing hippocampal memory deficits and mitochondrial dysfunction in transgenic AD models (Corona et al. 2010), and lowering plaque load and inflammation (Vilella et al. 2018). Methods for the delivery of zinc into the brain to achieve these beneficial effects are being developed. Indeed, a study by (Chhabra et al. 2015) investigated the use of injected nanoparticles loaded with zinc to target the CNS and enrich Zn^{2+} levels in the brain. This highly targeted drug delivery technique could significantly boost Zn^{2+} levels *in vivo* without resorting to more invasive methods of intracranial injection. It is an alternative to traditional orally administered zinc supplements and has the benefit of being precisely targeted to the brain, and bypasses issues such as inadequate drug absorption through the blood-brain barrier.

Over 16 studies since 1952 have determined that individuals with prostate cancer, another age-related disease, possess significantly lower serum zinc levels than men with normal and benign prostate tissue (Costello and Franklin 2016). All these studies consistently reported a reduction in mean zinc levels by as much as 80% in prostate cancer samples, potentially due to zinc's role as an inducer of apoptosis and regulator of citrate levels in the prostate. These processes prevent malignant cell proliferation in the prostate leading to prostate cancer (PCa). Indeed, a study of prostate cancer patients found that PCa and advanced disease severity was associated with lower zinc status (Wakwe et al. 2019). Boosting cellular zinc levels could prevent premalignant tissue from developing by promoting apoptosis and inhibiting cell proliferation and migration.

Considerations must be made for the optimal type of zinc supplement. Organic sources such as gluconates and citrates are preferable over inorganic sources of zinc owing to their higher bioavailability and better absorption (Sauer et al. 2017). Supplements like ZnAAs have added benefits for elderly patients in that medications may be less likely to interfere with them, besides compatibility with restricted dietary habits low in zinc.

Finally, herpes genitalis is a sexually transmitted infection caused by herpes simplex virus 1 and 2 (HSV-1 and 2). Topical use of zinc acetate gel has been proven effective in blocking the sexual transmission of herpes simplex virus-2 infection due to its microbicidal activity (Fernandez-Romero et al. 2012). However, this treatment approach also showed low toxicity *in vitro*, a common concern with the use of zinc salts. Also, a recent study has demonstrated the potential of a live virus vaccine using zinc oxide tetrapod nanoparticles (ZOTEN) and HSV-2 virus to prevent infection by HSV-2 (Agelidis et al. 2019).

Table 3. Beneficial effect of zinc supplementation in human health and disease

Clinical studies	Disease under investigation	Dosage of zinc supplement	Beneficial effects observed
Human			
Ranasinghe et al. 2018	Type 2 diabetes	20 mg Zn	Reduced blood glucose and insulin resistance
Eby 1984	Respiratory illness	23 mg Zn gluconate	Reduction in duration of the common cold
Barnes and Moynahan 1973	Acrodermatitis enteropathica	3 mg elemental Zn	Healing of skin lesions, regrowth of hair.
Saad et al. 2015	Epilepsy	0.2 g elemental Zn	Significant reduction of seizure rate
Mouse models			
Fourie et al. 2018	ASD	150 ppm Zn supplemented diet	Prevention of ASD-related behaviors
Cope et al. 2012	TBI-induced depression	180 ppm Zn supplemented diet + early intraperitoneal Zn injection	Significant reduction in anhedonia
Chen et al. 2000	Type II diabetes	20 ppm Zn supplemented diet	Improved hyperglycemia status in male mice

Conclusion

The trace metal zinc has diverse roles in physiological functioning and maintenance of good health. Consequently, zinc deficiency has repeatedly been associated with various chronic diseases that affect different systems, including neurological, respiratory, dermatological, and metabolic diseases. In particular, zinc deficiency has been implicated in many diseases of the CNS. Clinical studies involving both animal and human subjects reveal that dietary supplementation of zinc can reduce disease burden and restore good health in many cases. However, further research needs to be conducted on a larger scale to firmly establish zinc as a nutraceutical and identify zinc sources and supplements with maximal bioavailability. Zinc supplements paired with conventional therapies for mental disorders could significantly reduce symptoms and improve mental health. Importantly, not all zinc sources are created equal, with organic forms such as gluconates and amino acid chelates reported having higher bioavailability (Zhang et al. 2018). However, a more targeted delivery will be needed to explore the full potential of zinc as a nutraceutical. Future studies should focus on determining information such as optimal dosage and method delivery into the body.

References

Adamo, A.M., and Oteiza, P.I. (2010). 'Zinc deficiency and neurodevelopment: The case of neurons', *BioFactors* 36, 117–124. available: https://dx.doi.org/10.1002/biof.91.

Adams, J.B., Bhargava, A., Coleman, D.M., Frye, R.E. and Rossignol, D.A. (2021) 'Ratings of the Effectiveness of Nutraceuticals for Autism Spectrum Disorders: Results of a National Survey', *J Pers Med*, 11(9), available: https://dx.doi.org/10.3390/jpm11090878.

Agelidis, A., Koujah, L., Suryawanshi, R., Yadavalli, T., Mishra, Y.K., Adelung, R. and Shukla, D. (2019) 'An Intra-Vaginal Zinc Oxide Tetrapod Nanoparticles (ZOTEN) and Genital Herpesvirus Cocktail Can Provide a Novel Platform for Live Virus Vaccine', *Front Immunol*, 10, 500, available: https://dx.doi.org/10.3389/fimmu.2019.00500.

Akhondzadeh, S., Mohammadi, M.-R., and Khademi, M. (2004). 'Zinc sulfate as an adjunct to methylphenidate for the treatment of attention deficit hyperactivity disorder in children: A double blind and randomized trial [ISRCTN64132371]', *BMC Psychiatry* 4, 9. available: https://dx.doi.org/10.1186/1471-244X-4-9.

Ali, N., Fariha, K.A., Islam, F., Mohanto, N.C., Ahmad, I., Hosen, M.J. and Ahmed, S. (2021) 'Assessment of the role of zinc in the prevention of COVID-19 infections and mortality: A retrospective study in the Asian and European population', *J Med Virol*, 93(7), 4326-4333, available: https://dx.doi.org/10.1002/jmv.26932.

Ao, T., Pierce, J.L., Power, R., Pescatore, A.J., Cantor, A.H., Dawson, K.A., Ford, M.J. (2009) 'Effects of feeding different forms of zinc and copper on the performance and tissue mineral content of chicks', *Poult Sci*; 88 (10): 2171-5. available: https://dx.doi.org/10.3382/ps.2009-00117

Arens, M. and Travis, S. (2000) 'Zinc salts inactivate clinical isolates of herpes simplex virus in vitro', *J Clin Microbiol*, 38(5), 1758-62, available: https://dx.doi.org/10.1128/JCM.38.5.1758-1762.2000.

Bae, Y.S., Hill, N.D., Bibi, Y., Dreiher, J. and Cohen, A.D. (2010) 'Innovative uses for zinc in dermatology', *Dermatol Clin*, 28(3), 587-97, available: https://dx.doi.org/10.1016/j.det.2010.03.006.

Bagherani, N. and R Smoller, B. (2016) 'An overview of zinc and its importance in dermatology- Part I: Importance and function of zinc in human beings', *Global Dermatology*, 3(5), 330-336, available: https://dx.doi.org/10. 15761/god.1000185.

Bajait, C., Thawani, V., (2011). 'Role of zinc in pediatric diarrhea', *Indian J. Pharmacol.* 43, 232–235. https://doi.org/10.4103/0253-7613.81495

Banupriya, N., Bhat, B.V., Vickneshwaran, V., and Sridhar, M.G. (2020). 'Effect of zinc supplementation on relative expression of immune response genes in neonates with sepsis: A preliminary study', *Indian J. Med. Res.* 152, 296–302. https://doi:10.4103/ijmr.IJMR_557_18.

Bao, B., Prasad, A.S., Beck, F.W.J., Snell, D., Suneja, A., Sarkar, F.H., et al. (2008). 'Zinc supplementation decreases oxidative stress, incidence of infection, and generation of inflammatory cytokines in sickle cell disease patients', *Transl. Res.* 152, 67–80. https://doi:10.1016/j.trsl.2008.06.001.

Bao, S. and Knoell, D.L. (2006) 'Zinc modulates cytokine-induced lung epithelial cell barrier permeability', *Am J Physiol Lung Cell Mol Physiol*, 291(6), L1132-41, available: https://dx.doi.org/10.1152/ajplung.00207.2006.

Barnes, P.M. and Moynahan, E.J. (1973) 'Zinc deficiency in acrodermatitis enteropathica: multiple dietary intolerance treated with synthetic diet', *Proc R Soc Med*, 66(4), 327-9.

Baum, M.K., Campa, A., Lai, S., Lai, H. and Page, J.B. (2003) 'Zinc status in human immunodeficiency virus type 1 infection and illicit drug use', *Clin Infect Dis*, 37 Suppl 2, S117-23, available: https://dx.doi.org/10.1086 /375875.

Baum, M.K., Shor-Posner, G. and Campa, A. (2000) 'Zinc status in human immunodeficiency virus infection', *J Nutr*, 130 (5S Suppl), 1421S-3S, available: https://dx.doi.org/10.1093/jn/130.5.1421S.

Behjatian Esfahani, M.; Moravej, H.; Ghaffarzadeh, M.; Nehzati Paghaleh, G.A. (2021). 'Comparison of the Zn-Threonine, Zn-Methionine, and Zn Oxide on Performance, Egg Quality, Zn Bioavailability, and Zn Content in Egg and Excreta of Laying Hens', *Biol. Trace Elem Res* 2021; 199 (1): 292-304. available: https://dx.doi.org/10.1007/s12011-020-02141-8

Besag, F.M. (2018) 'Epilepsy in patients with autism: links, risks and treatment challenges', *Neuropsychiatr Dis Treat*, 14, 1-10, available: https://dx.doi.org/10.2147/NDT.S120509.

Bhalla, P., Chadha, V.D., Dhar, R. and Dhawan, D.K. (2007) 'Neuroprotective effects of zinc on antioxidant defense system in lithium treated rat brain', *Indian J Exp Biol*, 45(11), 954-8.

Bhatnagar, S. and Taneja, S. (2001) 'Zinc and cognitive development', *Br J Nutr*, 85 Suppl 2, S139-45, available: https://dx.doi.org/10.1079/bjn2000306.

Black, R.E. (2003) 'Zinc deficiency, infectious disease and mortality in the developing world', *J Nutr*, 133(5 Suppl 1), 1485S-9S, available: https://dx.doi.org/10.1093/jn/133.5.1485S.

Buff, C.E., Bollinger, D.W., Ellersieck, M.R., Brommelsiek, W.A., Veum, T.L. (2005) 'Comparison of growth performance and zinc absorption, retention, and excretion in weanling pigs fed diets supplemented with zinc-polysaccharide or zinc oxide', *J Anim Sci*, 83 (10): 2380-6. available: https://dx.doi.org/10.2527/2005.83102380x.

Cao, J., Henry, P.R., Guo, R., Holwerda, R.A., Toth, J.P., Littell, R.C., Miles, R.D., Ammerman, C.B. (2000). 'Chemical characteristics and relative bioavailability of supplemental organic zinc sources for poultry and ruminants', *J Anim Sci;* 78 (8): 2039-54. available: https://dx.doi.org/10.2527/2000.7882039x

Case, C.L., Carlson, M.S. (2002). 'Effect of feeding organic and inorganic sources of additional zinc on growth performance and zinc balance in nursery pigs', *J Anim Sci;* 80 (7): 1917-24. available: https://dx.doi.org/10.2527/ 2002.8071917x

Cezar, L.C., Kirsten, T.B., da Fonseca, C.C.N., de Lima, A.P.N., Bernardi, M.M., Felicio, L.F. (2018). 'Zinc as a therapy in a rat model of autism prenatally induced by valproic acid', *Prog Neuropsychopharmacol Biol Psychiatry*; 84 (Pt A): 173-180.

Chaidez, V., Hansen, R.L. and Hertz-Picciotto, I. (2014) 'Gastrointestinal problems in children with autism, developmental delays or typical development', *J Autism Dev Disord*, 44(5), 1117-27, available: https://dx.doi.org/10.1007/s10803-013-1973-x.

Chasapis, C.T., Loutsidou, A.C., Spiliopoulou, C.A. and Stefanidou, M.E. (2012) 'Zinc and human health: an update', *Arch Toxicol*, 86(4), 521-34, available: https://dx.doi.org/10.1007/s00204-011-0775-1.

Chasapis, C.T., Ntoupa, PS.A., Spiliopoulou, C.A. et al. (2020) Recent aspects of the effects of zinc on human health. *Arch Toxicol* 94, 1443–1460. https://doi.org/10.1007/s00204-020-02702-9

Chen, M., Sun, Y. and Wu, Y. (2020) 'Lower circulating zinc and selenium levels are associated with an increased risk of asthma: evidence from a meta-analysis', *Public Health Nutr*, 23(9), 1555-1562, available: https://dx.doi.org/10.1017/S1368980019003021.

Chen, M.D., Song, Y.M. and Lin, P.Y. (2000) 'Zinc effects on hyperglycemia and hypoleptinemia in streptozotocin-induced diabetic mice', *Horm Metab Res*, 32(3), 107-9, available: https://dx.doi.org/10.1055/s-2007-978600.

Chew, E.Y., Clemons, T.E., Sangiovanni, J.P., Danis, R.P., Ferris, F.L., 3rd, Elman, M.J., Antoszyk, A.N., Ruby, A.J., Orth, D., Bressler, S.B., Fish, G.E., Hubbard, G.B., Klein, M.L., Chandra, S.R., Blodi, B.A., Domalpally, A., Friberg, T., Wong, W.T., Rosenfeld, P.J., Agron, E., Toth, C.A., Bernstein, P.S. and Sperduto, R.D. (2014) 'Secondary analyses of the effects of lutein/zeaxanthin on age-related macular degeneration progression: AREDS2 report No. 3', *JAMA Ophthalmol*, 132(2), 142-9, available: https://dx.doi.org/10.1001/jamaophthalmol.2013.7376.

Chew, E.Y., *et. al.* (2013) 'Lutein + zeaxanthin and omega-3 fatty acids for age-related macular degeneration: the Age-Related Eye Disease Study 2 (AREDS2) randomized clinical trial', *JAMA*, 309(19), 2005-15, available: https://dx.doi.org/10.1001/ jama.2013.4997.

Chhabra, R., Ruozi, B., Vilella, A., Belletti, D., Mangus, K., Pfaender, S., Sarowar, T., Boeckers, T.M., Zoli, M., Forni, F., Vandelli, M.A., Tosi, G. and Grabrucker, A.M. (2015) 'Application of Polymeric Nanoparticles for CNS Targeted Zinc Delivery In Vivo', *CNS Neurol Disord Drug Targets*, 14(8), 1041-53, available: https://dx.doi.org/10.2174/1871527314666150821111 455.

Chukwuma, C.I., Mashele, S.S., Eze, K.C., Matowane, G.R., Islam, S.Md., Bonnet, S.L., Noreljaleel, A.E.M., Ramorobi, L.M. (2020). 'A comprehensive review on zinc(II) complexes as anti-diabetic agents: The advances, scientific gaps and prospects', *Pharmacol. Res.* 155, 104744. available: https://doi.org/10.1016/j.phrs.2020.104744

Colombo, J., Zavaleta, N., Kannass, K.N., Lazarte, F., Albornoz, C., Kapa, L.L., et al. (2014). 'Zinc Supplementation Sustained Normative Neurodevelopment in a

Randomized, Controlled Trial of Peruvian Infants Aged 6–18 Months', *The Journal of Nutrition* 144, 1298–1305. available: https://dx.doi. org/0.3945/jn.113.189365.

Cope, E.C., Morris, D.R., Gower-Winter, S.D., Brownstein, N.C. and Levenson, C.W. (2016) 'Effect of zinc supplementation on neuronal precursor proliferation in the rat hippocampus after traumatic brain injury', *Exp Neurol*, 279, 96-103, available: https://dx.doi.org/10.1016/j.expneurol.2016.02.017.

Cope, E.C., Morris, D.R., Scrimgeour, A.G. and Levenson, C.W. (2012) 'Use of zinc as a treatment for traumatic brain injury in the rat: effects on cognitive and behavioral outcomes', *Neurorehabil Neural Repair*, 26(7), 907-13, available: https://dx.doi.org/10.1177/1545968311435337.

Corona, C., Masciopinto, F., Silvestri, E., Viscovo, A.D., Lattanzio, R., Sorda, R.L., Ciavardelli, D., Goglia, F., Piantelli, M., Canzoniero, L.M. and Sensi, S.L. (2010) 'Dietary zinc supplementation of 3xTg-AD mice increases BDNF levels and prevents cognitive deficits as well as mitochondrial dysfunction', *Cell Death Dis*, 1, e91, available: https://dx.doi.org/10.1038/cddis.2010.73.

Costagliola, G., Nuzzi, G., Spada, E., Comberiati, P., Verduci, E., Peroni, D.G., (2021). 'Nutraceuticals in Viral Infections: An Overview of the Immunomodulating Properties', *Nutrients* 13, 2410. available: https://doi.org/10.3390/nu13072410

Costello, L.C. and Franklin, R.B. (2016) 'A comprehensive review of the role of zinc in normal prostate function and metabolism; and its implications in prostate cancer', *Arch Biochem Biophys*, 611, 100-112, available: https://dx.doi.org/10.1016/j.abb.2016.04.014.

Cousins, R.J. (2010) 'Gastrointestinal factors influencing zinc absorption and homeostasis', *Int J Vitam Nutr Res*, 80(4-5), 243-8, available: https://dx.doi.org/10.1024/0300-9831/a000030.

Daini, E., Hagmeyer, S., De Benedictis, C.A., Cristóvão, J.S., Bodria, M., Ross, A.M., Raab, A., Boeckers, T.M., Feldmann, J., Gomes, C.M., Zoli, M., Vilella, A., Grabrucker, A.M. (2021). 'S100B dysregulation during brain development affects synaptic SHANK protein networks via alteration of zinc homeostasis', *Transl Psychiatry;* 11 (1): 562.

Devaux, C.A., Rolain, J.M. and Raoult, D. (2020) 'ACE2 receptor polymorphism: Susceptibility to SARS-CoV-2, hypertension, multi-organ failure, and COVID-19 disease outcome', *J Microbiol Immunol Infect*, 53(3), 425-435, available: https://dx.doi.org/10.1016/j.jmii.2020.04.015.

'Dietary Reference Values for nutrients Summary report', (2017). *EFSA Supporting Publications*, 14(12), available: https://dx.doi.org/10.2903/sp. efsa.2017.e15121.

Du, Z., Hemken, R.W., Jackson, J.A., Trammell, D.S. (1996). 'Utilization of copper in copper proteinate, copper lysine, and cupric sulfate using the rat as an experimental model', *J Anim Sci;* 74 (7): 1657-63, available: https://dx.doi.org/10.2527/1996.7471657x

Eby, G.A., Davis, D.R. and Halcomb, W.W. (1984) 'Reduction in duration of common colds by zinc gluconate lozenges in a double-blind study', *Antimicrob Agents Chemother*, 25(1), 20-4, available: https://dx.doi.org/10.1128/AAC.25.1.20.

Fernandez-Romero, J.A., Abraham, C.J., Rodriguez, A., Kizima, L., Jean-Pierre, N., Menon, R., Begay, O., Seidor, S., Ford, B.E., Gil, P.I., Peters, J., Katz, D., Robbiani,

M. and Zydowsky, T.M. (2012) 'Zinc acetate/carrageenan gels exhibit potent activity in vivo against high-dose herpes simplex virus 2 vaginal and rectal challenge', *Antimicrob Agents Chemother*, 56(1), 358-68, available: https://dx.doi.org/10.1128/AAC.05461-11.

Fourie, C., Vyas, Y., Lee, K., Jung, Y., Garner, C.C. and Montgomery, J.M. (2018) 'Dietary Zinc Supplementation Prevents Autism Related Behaviors and Striatal Synaptic Dysfunction in Shank3 Exon 13-16 Mutant Mice', *Front Cell Neurosci*, 12, 374, available: https://dx.doi.org/10.3389/fncel. 2018.00374.

Ghishan, F.K., Said, H.M., Wilson, P.C., Murrell, J.E. and Greene, H.L. (1986) 'Intestinal transport of zinc and folic acid: a mutual inhibitory effect', *Am J Clin Nutr*, 43(2), 258-62, available: https://dx.doi.org/10.1093/ajcn/43.2.258.

Gibson, R.S. (1994) 'Zinc nutrition in developing countries', *Nutr Res Rev*, 7(1), 151-73, available: https://dx.doi.org/10.1079/NRR19940010.

Gibson, R.S., Raboy, V. and King, J.C. (2018) 'Implications of phytate in plant-based foods for iron and zinc bioavailability, setting dietary requirements, and formulating programs and policies', *Nutr Rev*, 76(11), 793-804, available: https://dx.doi.org/10.1093/nutrit/nuy028.

Gower-Winter, S.D. and Levenson, C.W. (2012) 'Zinc in the central nervous system: From molecules to behavior', *Biofactors*, 38(3), 186-93, available: https://dx.doi.org/10.1002/biof.1012.

Grabrucker, A.M. (2012) 'Environmental factors in autism', *Front Psychiatry*, 3, 118, available: https://dx.doi.org/10.3389/fpsyt.2012.00118.

Grabrucker, A.M., Knight, M.J., Proepper, C., Bockmann, J., Joubert, M., Rowan, M., Nienhaus, G.U., Garner, C.C., Bowie, J.U., Kreutz, M.R., Gundelfinger, E.D. and Boeckers, T.M. (2011) 'Concerted action of zinc and ProSAP/Shank in synaptogenesis and synapse maturation', *EMBO J*, 30(3), 569-81, available: https://dx.doi.org/10.1038/emboj.2010.336.

Grabrucker, S., Jannetti, L., Eckert, M., Gaub, S., Chhabra, R., Pfaender, S., Mangus, K., Reddy, P.P., Rankovic, V., Schmeisser, M.J., Kreutz, M.R., Ehret, G., Boeckers, T.M., Grabrucker, A.M. (2014). 'Zinc deficiency dysregulates the synaptic ProSAP/Shank scaffold and might contribute to autism spectrum disorders', *Brain;* 137 (Pt 1): 137-52. available: https://dx.doi.org/10.1093/brain/awt303

Grabrucker, S., Boeckers, T.M. and Grabrucker, A.M. (2016) 'Gender Dependent Evaluation of Autism like Behavior in Mice Exposed to Prenatal Zinc Deficiency', *Front Behav Neurosci*, 10, 37, available: https://dx.doi.org/10.3389/fnbeh.2016.00037.

Grabrucker, S., Haderspeck, J.C., Sauer, A.K., Kittelberger, N., Asoglu, H., Abaei, A., Rasche, V., Schon, M., Boeckers, T.M. and Grabrucker, A.M. (2017) 'Brain Lateralization in Mice Is Associated with Zinc Signaling and Altered in Prenatal Zinc Deficient Mice That Display Features of Autism Spectrum Disorder', *Front Mol Neurosci*, 10, 450, available: https://dx.doi.org/10.3389/ fnmol.2017.00450.

Gronli, O., Kvamme, J.M., Friborg, O. and Wynn, R. (2013) 'Zinc deficiency is common in several psychiatric disorders', *PLoS One*, 8(12), e82793, available: https://dx.doi.org/10.1371/journal.pone.0082793.

Guo, C.-H., and Wang, C.-L. (2012). 'Effects of Zinc Supplementation on Plasma Copper/Zinc Ratios, Oxidative Stress, and Immunological Status in Hemodialysis Patients', *Int. J. Med. Sci.* 10, 79–89. available: https://dx.doi.org/10.7150/ijms.5291.

Gupta, M., Mahajan, V.K., Mehta, K.S. and Chauhan, P.S. (2014) 'Zinc therapy in dermatology: a review', *Dermatol Res Pract*, 2014, 709152, available: https://dx.doi.org/10.1155/2014/709152.

Hagmeyer, S., Sauer, A.K. and Grabrucker, A.M. (2018) 'Prospects of Zinc Supplementation in Autism Spectrum Disorders and Shankopathies Such as Phelan McDermid Syndrome', *Front Synaptic Neurosci*, 10, 11, available: https://dx.doi.org/10.3389/fnsyn.2018.00011.

Hambidge M. 'Human zinc deficiency', *J Nutr.* 2000 May;130(5S Suppl):1344S-9S. available: https://dx.doi.org/10.1093/jn/130.5.1344S.

Hernandez-Camacho, J.D., Vicente-Garcia, C., Parsons, D.S. and Navas-Enamorado, I. (2020) 'Zinc at the crossroads of exercise and proteostasis', *Redox Biol*, 35, 101529, available: https://dx.doi.org/10.1016/j.redox.2020. 101529.

Hess, S.Y., Peerson, J.M., King, J.C. and Brown, K.H. (2007) 'Use of serum zinc concentration as an indicator of population zinc status', *Food Nutr Bull*, 28(3 Suppl), S403-29, available: https://dx.doi.org/10.1177/15648265070283 S303.

Islam, M.R., Attia, J., Ali, L., McEvoy, M., Selim, S., Sibbritt, D., Akhter, A., Akter, S., Peel, R., Faruque, O., Mona, T., Lona, H., Milton, A.H. (2016) 'Zinc supplementation for improving glucose handling in pre-diabetes: A double blind randomized placebo controlled pilot study', *Diabetes Res. Clin. Pract.* 115, 39–46. https://doi.org/10.1016/j.diabres.2016.03.010

Jackson, M.J. (1989) Physiology of zinc: general aspects. In: *Zinc in Human Biology*, edited by Mills CF. New York: Springer, p. 1–14.

Jayawardena, R., Ranasinghe, P., Galappatthy, P., Malkanthi, R., Constantine, G. and Katulanda, P. (2012) 'Effects of zinc supplementation on diabetes mellitus: a systematic review and meta-analysis', *Diabetol Metab Syndr*, 4(1), 13, available: https://dx.doi.org/10.1186/1758-5996-4-13.

Kelishadi, R., Hashemipour, M., Adeli, K., Tavakoli, N., Movahedian-Attar, A., Shapouri, J., et al. (2010). 'Effect of Zinc Supplementation on Markers of Insulin Resistance, Oxidative Stress, and Inflammation among Prepubescent Children with Metabolic Syndrome', *Metab. Syndr. Relat. Disord.* 8, 505–510. available: https://dx.doi.org/10.1089/met.2010.0020.

Kirby, K., Floriani, V., and Bernstein, H. (2001). 'Diagnosis and management of attention-deficit/hyperactivity disorder in children', *Curr. Opin. Pediatr.* 13, 190–199. available: https://dx.doi.org/10.1097/00008480-200104000-00019.

Kilic, M., Taskesen, M., Coskun, T., Gurakan, F., Tokatli, A., Sivri, H.S., Dursun, A., Schmitt, S. and Kury, S. (2012) 'A Zinc Sulphate Resistant Acrodermatitis Enteropathica Patient with a Novel Mutation in SLC39A4 Gene', *JIMD Rep*, 2, 25-8, available: https://dx.doi.org/10.1007/8904_ 2011_38.

King, J.C., Brown, K.H., Gibson, R.S., Krebs, N.F., Lowe, N.M., Siekmann, J.H. and Raiten, D.J. (2015) 'Biomarkers of Nutrition for Development (BOND)-Zinc Review', *J Nutr*, 146(4), 858S-885S, available: https://dx.doi.org/10.3945/jn. 115.220079.

Kirkil, G., Hamdi Muz, M., Seckin, D., Sahin, K. and Kucuk, O. (2008) 'Antioxidant effect of zinc picolinate in patients with chronic obstructive pulmonary disease', *Respir Med*, 102(6), 840-4, available: https://dx.doi.org/10.1016/j.rmed.2008.01.010.

Kirsten, T.B., Queiroz-Hazarbassanov, N., Bernardi, M.M., Felicio, L.F. (2015). 'Prenatal zinc prevents communication impairments and BDNF disturbance in a rat model of autism induced by prenatal lipopolysaccharide exposure', *Life Sci*; 130: 12-7.

Krebs, N.F., (2013). 'Update on Zinc Deficiency and Excess in Clinical Pediatric Practice', *Ann. Nutr. Metab.* 62, 19–29. https://doi.org/10.1159/000348261

Lai, J., Moxey, A., Nowak, G., Vashum, K., Bailey, K. and McEvoy, M. (2012) 'The efficacy of zinc supplementation in depression: systematic review of randomised controlled trials', *J Affect Disord*, 136(1-2), e31-e39, available: https://dx.doi.org/10.1016/j.jad.2011.06.022.

Lee, E.J., Lee, H., Huang, T.N., Chung, C., Shin, W., Kim, K., Koh, J.Y., Hsueh, Y.P., Kim, E. (2015). 'Trans-synaptic zinc mobilization improves social interaction in two mouse models of autism through NMDAR activation', *Nat Commun*; 6: 7168.

Lee, S.R. (2018) 'Critical Role of Zinc as Either an Antioxidant or a Prooxidant in Cellular Systems', *Oxid Med Cell Longev*, 2018, 9156285, available: https://dx.doi.org/10.1155/2018/9156285.

Li, L., Li, H., Zhou, W., Feng, J., Zou, X. (2019). 'Effects of zinc methionine supplementation on laying performance, zinc status, intestinal morphology, and expressions of zinc transporters' mRNA in laying hens', *J Sci Food Agric;* 99 (14): 6582-6588. available: https://dx.doi.org/10.1002/jsfa.9941

Liu, F.F., Azad, M.A.K., Li, Z.H., Li, J., Mo, K.B., Ni, H.J. (2020). 'Zinc Supplementation Forms Influenced Zinc Absorption and Accumu-lation in Piglets', *Animals (Basel)*; 11 (1). available: https://dx.doi.org/10.3390/ani11010036

Lukacik, M., Thomas, R.L., Aranda, J.V., (2008). 'A Meta-analysis of the Effects of Oral Zinc in the Treatment of Acute and Persistent Diarrhea', *Pediatrics* 121, 326–336. available: https://doi.org/10.1542/peds.2007-0921

Maares, M. and Haase, H. (2020) 'A Guide to Human Zinc Absorption: General Overview and Recent Advances of In Vitro Intestinal Models', *Nutrients*, 12(3), available: https://dx.doi.org/10.3390/nu12030762.

MacDonald, R.S. (2000) 'The role of zinc in growth and cell proliferation', *J Nutr*, 130(5S Suppl), 1500S-8S, available: https://dx.doi.org/10.1093/jn/130.5.1500S.

Maret, W. and Sandstead, H.H. (2006) 'Zinc requirements and the risks and benefits of zinc supplementation', *J Trace Elem Med Biol*, 20(1), 3-18, available: https://dx.doi.org/10.1016/j.jtemb.2006.01.006.

O'Brien, K.O., Zavaleta, N., Caulfield, L.E., Wen, J. and Abrams, S.A. (2000) 'Prenatal iron supplements impair zinc absorption in pregnant Peruvian women', *J Nutr*, 130(9), 2251-5, available: https://dx.doi.org/10.1093/jn/130.9.2251.

Ota, E., Mori, R., Middleton, P., Tobe-Gai, R., Mahomed, K., Miyazaki, C. and Bhutta, Z.A. (2015) 'Zinc supplementation for improving pregnancy and infant outcome', *Cochrane Database Syst Rev*, 2015/05/01(2), CD000230, available: https://dx.doi.org/10.1002/14651858.CD000230.pub5.

Petrilli, M.A., Kranz, T.M., Kleinhaus, K., Joe, P., Getz, M., Johnson, P., Chao, M.V. and Malaspina, D. (2017) 'The Emerging Role for Zinc in Depression and Psychosis', *Front Pharmacol*, 8, 414, available: https://dx.doi.org/10.3389/fphar.2017.00414.

Pfaender, S., Sauer, A.K., Hagmeyer, S., Mangus, K., Linta, L., Liebau, S., Bockmann, J., Huguet, G., Bourgeron, T., Boeckers, T.M., Grabrucker, A.M. (2017). 'Zinc deficiency and low enterocyte zinc transporter expression in human patients with autism related mutations in SHANK3', *Sci Rep;* 7: 45190. available: https://dx.doi.org/10.1038/srep45190

Ranasinghe, P., Wathurapatha, W.S., Galappatthy, P., Katulanda, P., Jayawardena, R. and Constantine, G.R. (2018) 'Zinc supplementation in prediabetes: A randomized double-blind placebo-controlled clinical trial', *J Diabetes*, 10(5), 386-397, available: https://dx.doi.org/10.1111/1753-0407.12621.

Read, S.A., Obeid, S., Ahlenstiel, C. and Ahlenstiel, G. (2019) 'The Role of Zinc in Antiviral Immunity', *Adv Nutr*, 10(4), 696-710, available: https://dx.doi.org/10.1093/advances/nmz013.

Ren, P.; Chen, J.; Wedekind, K.; Hancock, D.; Vázquez-Añón, M. (2020). Interactive effects of zinc and copper sources and phytase on growth performance, mineral digestibility, bone mineral concentrations, oxidative status, and gut morphology in nursery pigs. *Transl. Anim. Sci.,* 4, txaa083.

Russo, A.J. (2010) 'Decreased Serum Hepatocyte Growth Factor (HGF) in Individuals with Bipolar Disorder Normalizes after Zinc and Anti-oxidant Therapy', *Nutr Metab Insights*, 3, 49-55, available: https://dx.doi.org/10.4137/NMI.S5528.

Russo, A.J. (2011) 'Decreased zinc and increased copper in individuals with anxiety', *Nutr Metab Insights*, 4, 1-5, available: https://dx.doi.org/10.4137/NMI.S6349.

Rutter, G.A. (2010). 'Think zinc: New roles for zinc in the control of insulin secretion', *Islets* 2, 49–50. available: https://dx.doi.org/10.4161/isl.2.1.10259.

Saad, K., El-Houfey, A.A., Abd El-Hamed, M.A., El-Asheer, O.M., Al-Atram, A.A. and Tawfeek, M.S. (2015) 'A randomized, double-blind, placebo-controlled clinical trial of the efficacy of treatment with zinc in children with intractable epilepsy', *Funct Neurol*, 30(3), 181-5, available: https://dx.doi.org/10.11138/fneur/2015.30.3.181.

Sanna, A., Firinu, D., Zavattari, P. and Valera, P. (2018) 'Zinc Status and Autoimmunity: A Systematic Review and Meta-Analysis', *Nutrients*, 10(1), available: https://dx.doi.org/10.3390/nu10010068.

Sauer, A.K., Pfaender, S., Hagmeyer, S., Tarana, L., Mattes, A.K., Briel, F., Kury, S., Boeckers, T.M. and Grabrucker, A.M. (2017) 'Characterization of zinc amino acid complexes for zinc delivery in vitro using Caco-2 cells and enterocytes from hiPSC', *Biometals*, 30(5), 643-661, available: https://dx.doi.org/10.1007/s10534-017-0033-y.

Sauer, A.K., Stanton, J.E., Hans, S., Grabrucker, A.M. Autism Spectrum Disorders: Etiology and Pathology. (2021). In: Grabrucker, A.M., editor. *Autism Spectrum Disorders* [Internet]. Brisbane (AU): Exon Publications; Chapter 1.

Sauer, A.K., Vela, H., Vela, G., Stark, P., Barrera-Juarez, E. and Grabrucker, A.M. (2020) 'Zinc Deficiency in Men Over 50 and Its Implications in Prostate Disorders', *Front Oncol*, 10, 1293, available: https://dx.doi.org/10.3389/fonc.2020.01293.

Schlegel, P., Windisch, W. (2006). 'Bioavailability of zinc glycinate in comparison with zinc sulphate in the presence of dietary phytate in an animal model with Zn labelled rats', *J Anim Physiol Anim Nutr (Berl)*; 90 (5-6): 216-22. available: https://dx.doi.org/10.1111/j.1439-0396.2005.00583.x

Schoen, M., Asoglu, H., Bauer, H.F., Muller, H.P., Abaei, A., Sauer, A.K., Zhang, R., Song, T.J., Bockmann, J., Kassubek, J., Rasche, V., Grabrucker, A.M. and Boeckers, T.M. (2019) 'Shank3 Transgenic and Prenatal Zinc-Deficient Autism Mouse Models Show Convergent and Individual Alterations of Brain Structures in MRI', *Frontiers in Neural Circuits*, 13, available: https://dx.doi.org/ARTN 10.3389/fncir.2019.00006.

Silverberg, N.D. and Panenka, W.J. (2019) 'Antidepressants for depression after concussion and traumatic brain injury are still best practice', *BMC Psychiatry*, 19(1), 100, available: https://dx.doi.org/10.1186/s12888-019-2076-9.

Stanton, J.E., Malijauskaite, S., McGourty, K. and Grabrucker, A.M. (2021) 'The Metallome as a Link Between the "Omes" in Autism Spectrum Disorders', *Front Mol Neurosci*, 14, 695873, available: https://dx.doi.org/10.3389/fnmol.2021.695873.

Suo, H.; Lu, L.; Zhang, L.; Zhang, X.; Li, H.; Lu, Y.; Luo, X. (2015). 'Relative bioavailability of zinc-methionine chelate for broilers fed a conventional corn-soybean meal diet', *Biol. Trace Elem. Res*, 165, 206–213.

Swardfager, W., Herrmann, N., Mazereeuw, G., Goldberger, K., Harimoto, T., Lanctôt, K.L., (2013). 'Zinc in Depression: A Meta-Analysis', *Biol. Psychiatry, Cortico-limbic Connectivity and Mood Disorder*s 74, 872–878. https://doi.org/10.1016/j.biopsych.2013.05.008

Szewczyk, B. (2013) 'Zinc homeostasis and neurodegenerative disorders', *Front Aging Neurosci*, 5, 33, available: https://dx.doi.org/10.3389/fnagi.2013.00033.

Szewczyk, B., Branski, P., Wieronska, J.M., Palucha, A., Pilc, A. and Nowak, G. (2002) 'Interaction of zinc with antidepressants in the forced swimming test in mice', *Pol J Pharmacol*, 54(6), 681-5.

Trivedi, S.S., Chudasama, R.K., Patel, N., (2009). 'Effect of Zinc Supplementation in Children with Acute Diarrhea: Randomized Double Blind Controlled Trial', *Gastroenterol. Res*. 2, 168–174. https://doi.org/10.4021/gr2009.06.1298

Vela, G., Stark, P., Socha, M., Sauer, A.K., Hagmeyer, S. and Grabrucker, A.M. (2015) 'Zinc in gut-brain interaction in autism and neurological disorders', *Neural Plast*, 2015, 972791, available: https://dx.doi.org/10.1155/2015/972791.

Vilella, A., Belletti, D., Sauer, A.K., Hagmeyer, S., Sarowar, T., Masoni, M., Stasiak, N., Mulvihill, J.J.E., Ruozi, B., Forni, F., Vandelli, M.A., Tosi, G., Zoli, M. and Grabrucker, A.M. (2018) 'Reduced plaque size and inflammation in the APP23 mouse model for Alzheimer's disease after chronic application of polymeric nanoparticles for CNS targeted zinc delivery', *J Trace Elem Med Biol*, 49, 210-221, available: https://dx.doi.org/10.1016/j.jtemb.2017.12.006.

Vishwanathan, R., Chung, M. and Johnson, E.J. (2013) 'A systematic review on zinc for the prevention and treatment of age-related macular degeneration', *Invest Ophthalmol Vis Sci*, 54(6), 3985-98, available: https://dx.doi.org/10.1167/iovs.12-11552.

Vyas, Y., Lee, K., Jung, Y., Montgomery, J.M. (2020) 'Influence of maternal zinc supplementation on the development of autism-associated behavioural and synaptic deficits in offspring Shank3-knockout mice', *Mol Brain;* 13 (1): 110.

Wakwe, V.C., Odum, E.P. and Amadi, C. (2019) 'The impact of plasma zinc status on the severity of prostate cancer disease', *Investig Clin Urol*, 60(3), 162-168, available: https://dx.doi.org/10.4111/icu.2019.60.3.162.

Wang, M.X., Win, S.S., and Pang, J. (2020). 'Zinc Supplementation Reduces Common Cold Duration among Healthy Adults: A Systematic Review of Randomized Controlled Trials with Micronutrients Supplementation', *Am. J. Trop. Med. Hyg.* 103, 86–99. available: https://dx.doi.org/10.4269/ajtmh.19-0718.

Wedekind, K.J., Baker, D.H. (1990). 'Zinc bioavailability in feed-grade sources of zinc', *J Anim Sci*; 68 (3): 684-9. available: https://dx.doi.org/10.2527/1990.683684x

Wedekind, K.J., Hortin, A.E., Baker, D.H. (1992). 'Methodology for assessing zinc bioavailability: efficacy estimates for zinc-methionine, zinc sulfate, and zinc oxide', *J Anim Sci;* 70 (1): 178-87. available: https://dx.doi.org/10.2527/1992.701178x

Weigand, E., Kirchgessner, M. [Absorbability of zinc from different compounds]. *Z Tierphysiol Tierernahr Futtermittelkd* 1979; 42 (3): 137-46.

Wessels, I., Rolles, B. and Rink, L. (2020) 'The Potential Impact of Zinc Supplementation on COVID-19 Pathogenesis', *Front Immunol*, 11, 1712, available: https://dx.doi.org/10.3389/fimmu.2020.01712.

Wilson, D., Varigos, G., Ackland, M.L., (2006). 'Apoptosis may underlie the pathology of zinc-deficient skin', *Immunol. Cell Biol.* 84, 28–37. https://doi.org/10.1111/j.1440-1711.2005.01391.x

Xie, Y., Zhang, Q., Wang, L., Wang, Y., Cheng, Z., Yang, Z., Yang, W. (2019). 'The Effects of Partially or Completely Substituted Dietary Zinc Sulfate by Lower Levels of Zinc Methionine on Growth Performance, Apparent Total Tract Digestibility, Immune Function, and Visceral Indices in Weaned Piglets', *Animals (Basel)*, 9, 236.

Yakoob, M.Y., Theodoratou, E., Jabeen, A., Imdad, A., Eisele, T.P., Ferguson, J., et al. (2011). 'Preventive zinc supplementation in developing countries: impact on mortality and morbidity due to diarrhea, pneumonia and malaria', *BMC Public Health* 11, S23. available: https://dx.doi.org/10.1186/1471-2458-11-S3-S23.

Yoshikawa, Y., Ueda, E., Miyake, H., Sakurai, H., Kojima, Y. (2001). 'Insulinomimetic bis(maltolato)zinc(II) complex: blood glucose normalizing effect in KK-A(y) mice with type 2 diabetes mellitus', *Biochem. Biophys. Res. Commun.* 281, 1190–1193. available: https://dx.doi.org/10.1006/bbrc.2001.4456.

Zhang, J., Yang, W., Piquemal, J.P. Ren, P. (2012) 'Modeling Structural Coordination and Ligand Binding in Zinc Proteins with a Polarizable Potential', *J Chem Theory Comput*, 8(4), 1314-1324, available: https://dx.doi.org/10.1021/ct200812y.

Zhang SQ, Yu XF, Zhang HB, Peng N, Chen ZX, Cheng Q, Zhang XL, Cheng SH, Zhang Y. (2018b). 'Comparison of the Oral Absorption, Distribution, Excretion, and Bioavailability of Zinc Sulfate, Zinc Gluconate, and Zinc-Enriched Yeast in Rats', *Mol Nutr Food Res*; 62 (7): e1700981. available: https://dx.doi.org/10.1002/mnfr.201700981

Zhang, Y., Ward, T.L., Ji, F., Peng, C., Zhu, L., Gong, L. and Dong, B. (2018) 'Effects of zinc sources and levels of zinc amino acid complex on growth performance, hematological and biochemical parameters in weanling pigs', *Asian-Australas J Anim Sci*, 31(8), 1267-1274, available: https://dx.doi.org/10.5713/ajas.17.0739.

Chapter 2

Impacts and Significance of Metal–Microbe Interactions on Soil and Sediment Ecosystems

Binu Prakash and Mahesh Mohan*
School of Environmental Sciences, Mahatma Gandhi University, Kottayam, Kerala, India

Abstract

Modern globalisation has escalated the sources of heavy metal pollution in human-centered natural habitats. Heavy metals are persistent and are toxic to all forms of life. So, metal detoxification and restoration of metal-contaminated sites are very significant. Detoxification with microbes is very relevant as it is a cost-effective and natural method. Microorganisms living in already contaminated environments are often well adapted to survive in the presence of existing contamination. Microorganisms like algae, bacteria, and fungi can detoxify trace metals by bioremediation. Many reports on metal-microbe interactions highlight its important role in eradicating heavy metals from the ecosystem in an eco-friendly way through removal, detoxification, and recovery of organic and inorganic metals. There is a long quandary on how microbes link with metals in both natural and manmade environmental or biogeochemical cycling of metals by microorganisms. This chapter, therefore, attempts to address the different aspects of metal-microbial interactions, focusing on soil and sediment ecosystems.

Keywords: heavy metal, bioremediation, microbes, microbial remediation, eco-genomics

---

* Corresponding Author's Email: maheshmohan@mgu.ac.in.

In: Trace Metals: Sources, Applications and Environmental Implications
Editor: Oscar M. Thygesen
ISBN: 978-1-68507-797-6

1. Introduction

Metals are double-edged swords for microorganisms, which remain as essential micronutrients at low concentrations, but are toxic above certain limits. Moreover, microorganisms are attached to various types of metals and metalloids in the environment. It interacts with metals according to whether the organism is a prokaryote or a eukaryote (Boteva et al., 2016). Microorganisms exposed to heavy metal loads can adapt to metal contamination (Munoz et al., 2006) and play a salient role in the environmental fate of toxic metals. In general, the strategy that microorganisms follow is to avoid the accumulation of excessive metal concentrations. High concentrations of heavy metals in the environment cause serious health problems owing to their non-degradability, which has long-term harmful effects on ecosystems (Singh et al., 2011). Roane et al. (2015) reported that microorganisms directly affect the fate of metals in the environment, hence, they can influence the extent of pollution. Several studies highlighted that only long-term exposure to heavy metal contamination leads to the development of adaptive microbial communities in the ecosystem (Tobor-Kapłon et al., 2005; Hoostal et al., 2008; Boteva et al., 2016). Microorganisms may not have a very specific uptake system for most metals. Especially, the surface properties of bacteria determine their metal adsorption properties. Possible cellular systems may be engrossed in microbial resistance and resistance to excessive concentrations of heavy metals in the environment.

Microbial populations have diverse characteristics that allow living in highly contaminated areas (Wei et al., 2014). They can survive and grow in these unfavourable ecosystems. The guaranteed property of these organisms is that they must have evolved an efficient enzymatic network and / or mechanism for uptake, accumulation, detoxification and conversion of toxic elements (Begum et al., 2021). Microbes participate in the mineralisation of some organic contaminants into end-products and utilise metabolic intermediates as their primary substrates for cell growth.

Almost all metal-microbial interactions have been studied in connection with environmental biotechnology, even as removing, recovering, or detoxifying inorganic and organometallic pollutants. The relatively sudden invasion of pollutants into the recipient ecosystem overwhelmed their self-cleaning ability and resulted in the accumulation of pollutants. Due to the toxicity and ubiquity of metals in the environment, microorganisms have developed several methods for treating metals with essential and unwanted toxicity. So, the current chapter focuses on specifically different aspects of metal-microbe interactions, including mechanisms of metal tolerance, interaction machinery, genomic mechanisms, along with its positive and negative effects.

2. Metal Contamination Overview

Metals are naturally occurring constituents in the environment and their concentrations vary by geographic region. The release of pollutants into the environment by human activity has increased significantly over the last few decades. One of the main problems is the degradation and urban soil contamination in many parts of the world. Urban soils undergo continuous accumulation of pollutants from local or diffuse sources. A metal or metalloid species can be considered as a pollutant if it occurs in an undesired place or in a form or concentration that has a deleterious effect on humans or the environment. The term 'heavy metal' refers to a metal element that is relatively dense, above 4 g / cm^3, and has toxic or toxic effects even at very low concentrations (Mishra et al., 2016). According to Singh et al. (2011), metals or metalloids include chromium, nickel, zinc, lead, arsenic, cadmium, mercury, selenium, silver and copper.

Pollution can be related to inapt changes in the physical, chemical, and biological properties of air, soil, and water harmful to living organisms, both animals and plants (Mishra et al., 2019). Metals are considered non-degradable pollutants instead of organic pollutants, which can be broken down into less harmful components by biological or chemical processes (Wang et al., 2002). On the other hand, pollution denotes the presence of higher amounts of substances in the environment that go beyond the natural background strength of the area and the organism. Metals are predominantly cation species, and metalloids are predominantly anionic species (Gomathy and Sabarinathan, 2010). Major industrial wastewater/waste containing heavy metals is a common source, including metal refining, nuclear and electronics, metal refining plants, and glass. Water contaminated with heavy metals is highly carcinogenic and is toxic even at relatively low concentrations. Metal pollution of land resources remains the focus of many environmental studies and is receiving a great deal of global attention.

Typical pollutants include persistent toxins such as trace metals and persistent organic pollutants discharged from the industry, emissions from transportation, and wastes from municipal activities (Wong et al., 2006; Fabietti et al., 2010). The anthropogenic tactics of heavy metals reportedly move past the herbal fluxes for a few metals. Metals emitted in wind-blown dust are ordinarily from commercial areas (He et al., 2005). Trace metal contamination in sediments helped to identify metal contamination in the past due to various anthropogenic impacts and changes in land-use patterns. Potentially harmful metals are major pollutants in urban road dust that reflect industry characteristics and environmental qualities.

Traces of metals and metalloids persistent in the environment are not biodegradable, not thermally degradable, and therefore easily accumulate at toxic concentrations. Trace metals enter the soil environment through both pedogenic and anthropogenic processes. Due to increased metal emissions, trace metal pollution is a global problem in marine environments (The Mermex Group et al., 2011). Most of these elements are found naturally in soil materials, with the primary source of being weathering of soil bed materials, including igneous and sedimentary rocks and coal. Metal smelting is one of the main anthropogenic sources of heavy metals and metalloid in the environment (Wang et al., 2013).

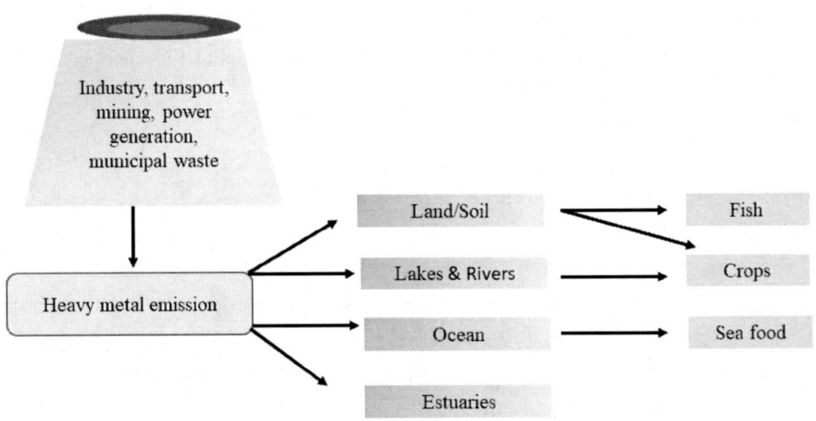

Figure 1. Fate of heavy metals in the environment.

3. Bioaccumulation and Toxicity

Bioaccumulation of chemicals is a problem for all living things. This results from a dynamic balance between ingestion and excretion, intake and elimination. The degree of bioaccumulation determines the toxic effects. Bioaccumulation refers to the way that contaminants enter a food chain and the subsequent accumulation of contaminants in biological tissue by aquatic organisms from sources such as water, food, and particulates. Trace elements can have toxic effects on plants and humans. Metals are unique as they are not subject to potentially variable chemical or biologically induced degradation or reduced toxicity over time. Therefore, countries and organisations establish maximum regulatory limits for their concentrations in soils. These limit values are generally divided into different categories depending on the use and properties of the soil (Robinson, 2009). Toxic pollutants come from diverse sources, such as industrial sewage, acid rain, gold ore, and metal ions remaining in the soil. Kido (2013) stated that each metal has different concentrations that are considered noxious to both the environment and the human body. These toxic pollutants pose consequential health threats to humans, including bone loss, hepatotoxicity, skin problems and kidney damage.

Most heavy metals are required in low concentrations in biological systems for various metabolic processes, while high concentrations are required in biological systems, including humans and animals (Roane, 2015). Bioaccumulation is an active process that depends on the metabolic energy of various micro-level organisms. It is an energy-dependent heavy metal transport system. In addition, possible bioaccumulation methods or processes of heavy metal influx through the bacterial membrane include ion pumps, mediated transport, ion channels, endocytosis, both complex osmosis, and lipid osmosis. The bioaccumulation process is adversely affected by the low temperature in without energy sources and the exitances of metabolic inhibitors (Deng and Wang, 2012). Thus, the total accumulation of metal is determined by two cell properties: the sorption capacity of the cell envelope and the capability to absorb metals in the cytosol.

The metal binds to many cellular ligands through ionic interactions, moving essential metals away from the normal binding site. Metals additionally destroy proteins by attaching to sulphhydryl groups, and nucleic acids destroy by attaching to hydroxyl or phosphate groups (Tamas et al., 2014). Several studies (De Jonge et al., 2012; Mishra et al., 2016; Osman et al., 2019) reported that heavy metals could disrupt important enzyme function through competitive or non-competitive interactions with substrate, which leads to changes in enzyme composition. Heavy metals destroy metabolic (catabolic and anabolic) function in two ways:

I. They accumulate and consequently destroy the function of important glands and vital organs such as the heart, liver, brain, kidneys, and bones.

II. Transfer important food minerals from their original location and thus interfere with their biological function.

For example, cobalt is relatively destructive to most aquatic creatures. Prolonged exposure of fish to cobalt has toxicological impacts. These pernicious effects include decreased glycogen level in muscle, hyperlactic acidosis and necrosis of gill epithelial cells, and inadequate oxygen intake (Griffitt et al., 2008).

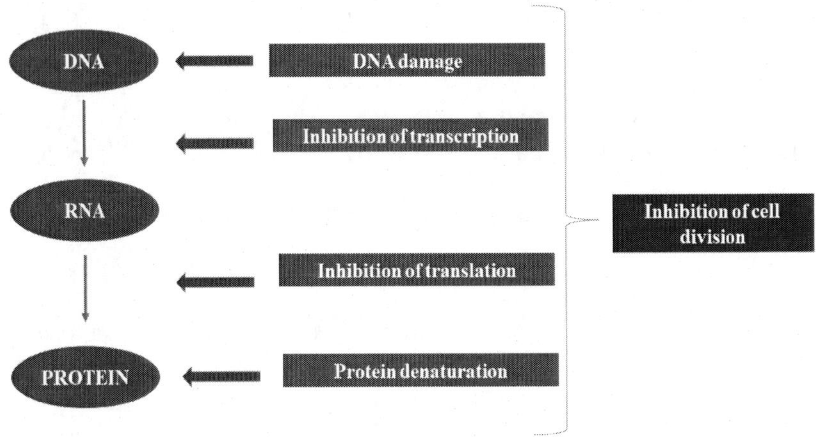

Figure 2. Summary of the various toxic influences of metals on the microbial cell.

4. Metal–Microbe Interactions

Microorganisms play important geological roles in the biosphere, especially in the fields of biogeochemical cycles of element, metal, mineral metabolism, and degradation (Reddy et al., 2012; Gadd, 2013). Unique microorganisms like bacteria and fungi serve an important role in converting heavy metals from one oxidised state to another. Contaminated soil, water, and other sources had an increased frequency of metal-resistant and resistant microorganisms (Issazadeh, et al., 2013; Yang et al., 2015). Microbial functions in soils are quite resistant to metal contamination, as they often retrieve after prime inhibition. According to Holtan-Hartwig et al. (2002), this is probably related to the fact that tolerant organisms replace sensitive ones, change the composition of the community, and increase their conditional tolerance to metals. Metal-resistant microorganisms with different phenotypic characteristics also affect heavy metals in the environment.

Microbes have several origins, which could impact adjustments in metallic speciation, toxicity and mobility, in addition to the mineral formation or mineral dissolution, or deterioration (Gadd et al., 2016). Metal–mineral–microbial interactions are central to the substructure of geomicrobiology and essential for the microbial biomineralisation process (Konhauser, 2007). Various studies revealed that (Bottjer, 2005; Chorover et al., 2007) ceratin types of microorganisms consist of both prokaryotes and eukaryotes. Their mutual and symbiotic relationships with higher organisms may actively exist put up to geological phenomena. Microorganisms can play a major role in the biogeochemical cycle of toxic heavy metals, including the purification of metal-contaminated substances, which are very important for use in repairs of metal-contaminated areas (Hemambika et al., 2011).

Table 1. Different bacterial surface components and metal binding functional groups

Functional Group	Location	Organism	Reference
Carboxyl	Lipopolysaccharide	*Pseudomonas aeruginosa*	Langley et al (1999)
Amine	Polypeptide	*Bacillus subtilis*	Beveridge and Murray (1980)
Thiol	Phytochelatins	*Escherichia coli*	Bae et al (2000)
Phosphoryl	Lipopolysaccharide	*Escherichia coli*	Ferris and Beveridge (1986)

Of course, microorganisms are involved in rock weathering, metal recruitment from minerals, metal reduction and oxidation, and metal precipitation and deposition. For mineral and metal compounds in waste, various metabolic processes (metal conversions) can occur when microbes interact with solid metals or metals in solution (Brandl, 2001). Microbes can dissolve or precipitate minerals and obtain energy from the oxidation/reduction process of metal ions. Its metabolic procedures are associated with the availability of metal ions and protects itself (cellular toxicity) from toxic amounts of metals through detoxification processes. Microorganisms can dissolve metal ions from minerals such as iron in pyrite and permuted them into soluble species (Gadd, 2010). The metabolic process works in reverse and can lead to the formation of insoluble species. Specific metal ions and their coordination complexes play various basic roles in the growth and metabolism of microorganisms. Other metal species are toxic to the life of microorganisms.

Bacteria have several mechanisms to resist heavy metals and detoxify them. Bacterial response to heavy metals depends on the category of bacteria and their inherent capability to withstand metal concentrations. More often, the bacterial reaction or response can be divided into two categories: general mechanisms that mediate resistance but are independent of metal stress for activation, and the second one is mechanisms that depend on activation by -specific metals (Osman et al., 2019). Examples of microbially important microbial communities directly intricated in geochemical transformation include manganese-oxidising, sulphate-reducing bacteria, iron-oxidising and -reducing bacteria, and reducing bacteria, sulphur-oxidising bacteria, etc (Gadd et al., 2000). Heavy metals can bind to the surface of microorganisms and penetrate cells (Yin et al., 2019). Several studies (Gadd, 2000; Chen et al., 2004) showed that, the iron and sulphur-oxidising bacteria *Thiobacillus ferrooxidans* and *Thiobacillus thiooxidans* were concentrated from polluted soil and were capable to leach more than 50 per cent of the existing metals.

In the process of evolution, microorganisms have modified the ability to live in almost all environmental conditions on earth or the ecosystem. These microorganisms can grow with or without oxygen or light, dissolve or precipitate ores and extract energy from the reduction/oxidation of metal ions (Raab and Feldmann, 2003). Metabolism is usually associated with the availability of essential amounts of metal ions and protects itself from toxic amounts of metal through the detoxification process. Conversion of metals and minerals by microbes can lead to deterioration and decay of synthetic and natural materials, mineral-based building materials, rock, acid mine spills, metal biocorrosion and related metal contamination (Gadd, 2013). Various microorganisms participate in the process of mineralisation. The formation of crystalline inorganic species under the influence of cells is called biomineralisation. Bone formation from calcium phosphate and formation of magnetite particles by magnetic bacteria belong to this category.

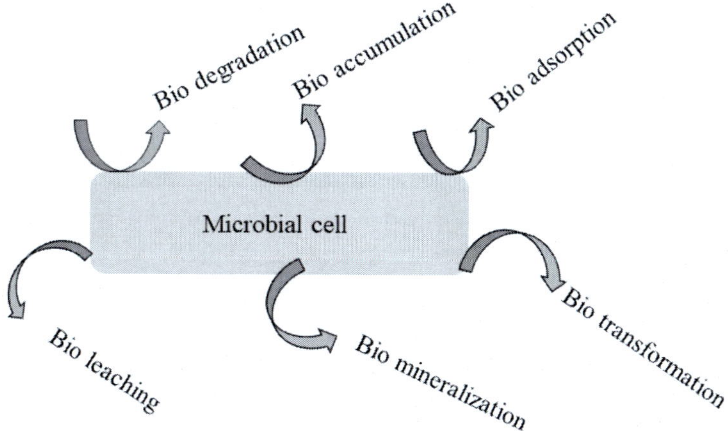

Figure 3. Different types of biodegradation methods by microbial cell.

5. Microbes and Metal Tolerance

Microbes have learned to live under aerobic or anaerobic conditions in extremely harsh circumstances (Hoehler and Jorgensen, 2013). Microbes can potentially fortify metals through an active, passive, or metabolic-dependent process, independent of metabolism. Xu et al. (2019) stated that heavy metal contamination in soil has higher energy requirements under metal stress and reduces microbial carbon consumption, resulting in significant changes in the microbial population. Thus, the total accumulation of metal is decided by two cellular properties: the sorption capacity of the cell envelope and the ability to take up metals in the cytosol compartment.

Soil microorganisms can improve phytoextraction procedure in several ways: they can upgrade the availability, solubility, and transport system of trace elements and nutrients by diminishing soil pH gradient, secretion of chelators and siderophores or redox changes; they can also enhance shoot and root growth, plant biomass production etc., (Becerra-Castro et al., 2009). Microbes can potentially accumulate metals through an independent metabolic passive process or an active metabolic active. Therefore, total accumulation is determined by two properties of the cell: the sorption capacity from the cell envelope and the metal uptake capacity in the cytosol (Haferburg & Kothe, 2007). Sprocati and their team (2006) explained that heavy metals act on microbial cells in different ways like macro and microscopic level alterations, bringing general changes in morphology side. The potential species like *Pseudomonas, Corneybacterium, Rhodoccocus, Bacillus, Alcaligenes, Candid*a, and *Cyberlindnera* can accumulate and sequester the intracellular metal ions (Moreno -Mateos et al. 2017; Chellaiah, 2018).

Usually, the species of Geobacter grows on the surface of iron ore. Iron is only available in direct contact with minerals (Tan et al., 2016). Microorganisms adapt to and resist heavy metals in heavily contaminated areas. The extracellular polymer substance present in the cell wall of biomass can attach to heavy metals via mechanisms such as proton exchange or metal microprecipitation. Microbe-metal resistance can be split up into three categories. These include: (a) General resistance mechanism that does not need metal stress (b) General resistance mechanism that is switched on by metal stress (c) Resistance mechanism which especially relies on a particular metal for their activation strategy (Orell et al., 2010; Roane et al., 2015). The interaction of these metals with microorganisms results in slow growth of unusual morphological changes and prohibition of biochemical processes in each cell. The virulent effects of metals are further seen at the community level (Wichard, 2015). Kushwaha and his team (2017) isolated the lead-resistant strain from the coal mine and identified it as *Acinetobacter junii* Pb1 by 16S rRNA sequencing. They also reported that these strains as well resistant other heavy metals such as copper, zinc, arsenic, chromium, mercury and cadmium. In 2020, Cyriaque and his team highlighted that bacterial interactions induced by metals promote diversity in the microbial flora of river sediments.

Fungus is well known for detoxifying heavy metals through active uptake, extracellular and intracellular precipitation strategy, and valence conversion. The recent report by Rahman et al. (2019) pointed to numerous species of fungi, which can incorporate certain heavy metals (Hg, Cd, Cu, Pb, and Zn) into the mycelium and spores. A large number of filamentous fungi come under the genera *Trichoderma, Penicillium, Aspergillus* and *Mucor*, which can withstand heavy metal stress (Vickers, 2017). Heavy metal contaminated by *Tricholomalobynsis* effect of metal solubilising microorganisms on the extraction of zinc and cadmium from soil (Maodaa et al., 2016).

6. Response of Microbial Communities on Different Metal Contamination

Interactions between metal and microorganisms during the metal desorption process on the cell surface through reduction or oxidation phenomena determines the fate of heavy metals in the environment. Some bacteria reduce toxic and mutagenic hexavalent chromium to less toxic trivalent chromium (Cheung et al., 2007). Various studies have recently been conducted on arsenic to explore the physiological and molecular mechanisms of toxicity, accumulation, detoxification, and tolerance of arsenic. Different microbial communities are associated with metal in diverse ways.

Bio-adsorbed arsenic ions and compounds (Jasrotia et al., 2014), with the help of sites strongly exposed to arsenic-treated with filamentous *Shiogusa* species, functional groups on the cell wall surface. Arsenic-resistant mushroom species can convert toxic metals into volatile/non-volatile products associated with microbes. In the case of copper, copper sulphide is oxidised to copper sulphate by microorganisms, and the aqueous phase contains metallic components. Lead is released in amazing amounts from industry and mining activities. Mercury is atypical in that it can also be volatilised by reduction. Mercury resistance or tolerance includes enzymatic reduction of mercury (Hg^{2+}) to the elemental mercury (Hg0) in both gram-positive and gram-negative bacteria (Nascimento et al., 2003; Sugio et al., 2009). For example, Hg (II) undergoes methylation by some bacteria, such as *Bacillus, Escherichia*, *Clostridium, and, Pseudomonas aeruginosa*, can be biomethylated into gaseous methylmercury.

Pushkar and his team (2019) isolated and identified mercury resistant bacteria such as *Bacillus sp. strain CSB_B078, Enterobacter sp. strain 08, Klebsiella pneumoniae isolate 23, Enterobacter sp. strain Amic_7, Klebsiella pneumoniae strain FY2*. These were isolated from both low and high-salinity areas of Mithi river, Mumbai, India. Catabolic Selenium (IV) reduction to biochemical Se (0) with chemical reducing agents such as sulphides and hydroxylamines, or with glutathione reductase, is the most important organism for the repair of Seoxyanions in anoxic deposits (Nejad et al., 2018). The *Klebsiella planticola* strain produces hydrogen sulphide from thiosulphate under anaerobic conditions and precipitates cadmium ions into their insoluble sulphides. Degradation of Cd (II) and Zn (II) is carried out according to the ion exchange process with *Saccharomyces cerevisiae* (Wang, 2012). Cr (VI) is toxic to most bacteria and mutagenic. This causes cells to grow, expand, inhibit cell division, and ultimately contribute to the inhibition of cell proliferation (Turpeinen et al., 2004). Iron-oxidising bacteria, such as members of the genera *Thiobacillus*, *Leptospirillum*, and *Ferroplasma*, use Fe^{2+} as an electron donor to meet their energy needs. Several bacteria, *Shewanellaalga, Acidithiobacillus thioxidans, Acidithiobacillus ferrooxidans*, and *Leptospirillum ferrooxidans*, have been reported to induce As (V) recruitment from minerals containing iron or sulphur (Turpeinen et al., 2004; Jasrotia et al., 2014).

6.1. Special Focus on Soil and Sediments

Healthy soil is important for the ecosystem. Pollution of aquatic ecosystems with trace metals poses notable health and environmental risk to invertebrates, fish, humans, and mangroves. The respective extent of the presence, persistence, different sources, nature of bioaccumulation and lethality of trace metals specifically in aquatic shores ecosystems are increasing globally (Maanan et al., 2013). The presence of heavy metals in soils is a serious problem due to their endurance in food chains and lack of biodegradability, which pulls down the entire ecosystem (Singh et al., 2011). Organic pollutants can be biodegradable, but their degradation rate is reduced by the availability of heavy metals in the environment. Elevated levels of trace metals in soil can impair microbial growth and enzymatic activity, slowing down biochemical processes in the soil environment.

Due to human activity, metal has accumulated in the soil. Such contaminated soil can form metal sinks from which surface water, groundwater, and vents can be contaminated (Bento et al., 2003). The toxic effects of mining wastes on human health worldwide are classified as acute, chronic, and extrinsic. They affect mainly the soil system, food chain, and groundwater properties. Adaptation to heavy metal-rich environments results in microorganisms that exhibit biosorption, precipitation, sequestration by extracellular ways, different transport mechanisms, and/or diverse chelation activities. The aquatic phase dispenses a transmit medium for metal transfer and circulation, organisms, and the aquatic environment through soil (Warren and Haack, 2001). The interaction between the metal mobilisation mechanism and the metal anchoring force is very complex and depends on the soil properties (Maanan et al., 2013).

Various research reports have shown microbial enzyme activity in sediments as an indicator of carbon and nutrient limitation. Sediments form an important sink for solid phase immobilisation and metal accumulation. Due to the widespread use and persistence of these metals, they are potentially toxic to the aquatic biota and affect their ecological function (Jaiswal and Pandey, 2018). Their bioavailability and concentration can lead to malignant effects on river biota, chiefly microbial activity in sediments. Moreover, heavy metals in deposits are often considered inert or stable in sediments. Sedimentary microbial communities are essential for nutrient cycle and organic matter remineralisation in coastal ecosystems. This is because, as the soil environment rises, they become mobile and can be released into the upper water column (Ke et al., 2017). Heavy metals can affect the metabolism of sediments. The measurement of sediment metabolism are an external or eco enzyme that breaks down organic matter into a soluble substrate for microbial assimilation. Microbial contributions to the solid phase distribution of metals in sediments include a series of sorption and precipitation reactions. In addition to this, Morel and Price (2003) stated that marine microorganisms, such as plankton-based microbes, play a vital role in the biogeochemical cycle of most marine essential metals. Jaiswal and Pandey (2018) reported various impacts of heavy metals on the reactivity of some microbial enzyme categories in the riverbed sediments in River Ganga, India.

Respirable anaerobic bacteria of organic halogenates are very important candidates for biological repair, as contaminated areas such as aquatic sediments, submerged soil, and groundwater are low in oxygen. The bioremediation process by microorganisms helps to reduce, remove, contain, or convert pollutants present in the soil, sediments, water, or air into harmless products. The passive entailment of microorganisms in these strategies, which are related to the behavior of individual microbial cells as an adsorbent of dissolved metal. Li and Ramakrishna (2011) studied different metal-resistant bacteria from lake sediments. Previously, eight copper-resistant strains were isolated from a copper mine from top soil-contaminated sediments of Lake Torch, Michigan.

7. Mechanism of Trace Metals Uptake by Microorganisms

The resistance to metals of microbes is achieved by intracellular and extracellular mechanisms - metals can be expelled through efflux transport strategies. Cytosol sequestering compounds can bind and detoxicate metals within the cell. Discharge of various chelators into the extracellular environment results in bound and attached metals (Lovley, 2003). The effects of microorganisms increase to a reasonable extent with increasing temperature, and improves microbial metabolism and enzymatic activity, accelerating biological repair. The stability of the microbial metal complex depends on the site of accretion, the cell wall composition of microbial cells, and the ionisation of chemicals on the cell wall (Tobor-Kaplon et al., 2005; Xu et al., 2019). The metabolic activity of some microorganisms indirectly causes metal precipitation through the formation of reactive inorganic ligands such as sulphide, phosphate, dissolved inorganic carbon (Ferris, 2000). Long-term contamination of heavy metals in soil is a critical problem that can promote bacterial species, which can develop heavy metal resistance. Moreover, the bioavailability of metals in habitats includes soil properties, climatic conditions and microbial activity. Certain members of the microbial community are more sensitive to heavy metal exposure than other members, depending on the sensitivity of important metabolic pathways. Another mechanism of microbial resistance to metals is the development of metal-resistant enzymatic forms, which is expected to be the major pathway for denitrifying bacteria, as metal pumps (Hinojosa et al., 2004).

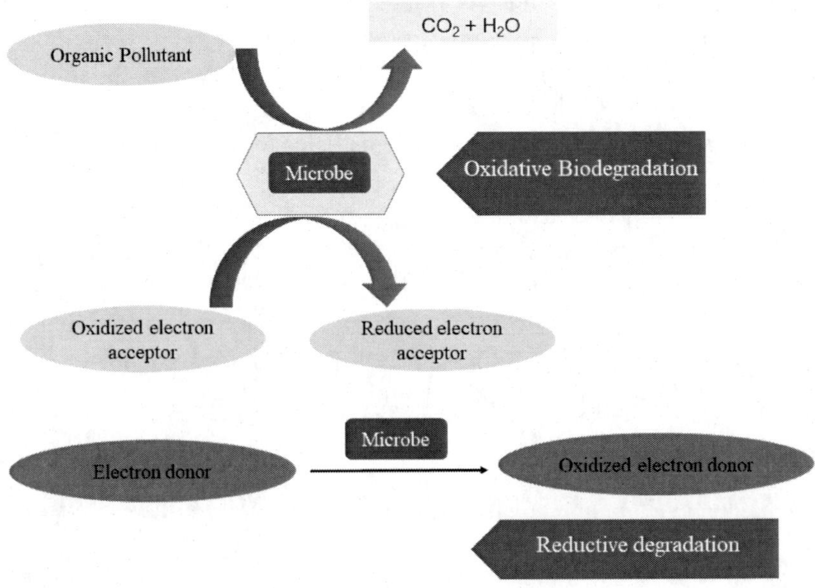

Figure 4. Oxidative and reductive degradation by microbes.

Microorganisms always have a negative charge on the surface of cells because of anionic structures that allow them to bind to metal cations. Cells have developed a series of mechanisms to cope with a deficiency or excess of metals (Hall, 2002). Primitive life forms such as archaea, both prokaryotic and eukaryotic bacteria, and fungi which are microbes, have developed a particularly rich variability of mechanisms to cope with intolerable or unusual environmental conditions. Microbes can convert metals to organometallic compounds both soluble and volatile, for example arsenic in to dimethylarsinic acid or trimethyl arsine. The contribution of these biologically produced or mobilised or biologically precipitated metals to the distribution of metals in the environment is as important as the pure physicochemical reactions themselves (Roane et al., 2015). Microorganisms frequently utilise specific routes to transport essential metals through the cell membrane to the cytoplasm. Unfortunately, harmful metals can as well penetrate membranes via diffusion, non-specific uptake systems, or pathways developed for other metals.

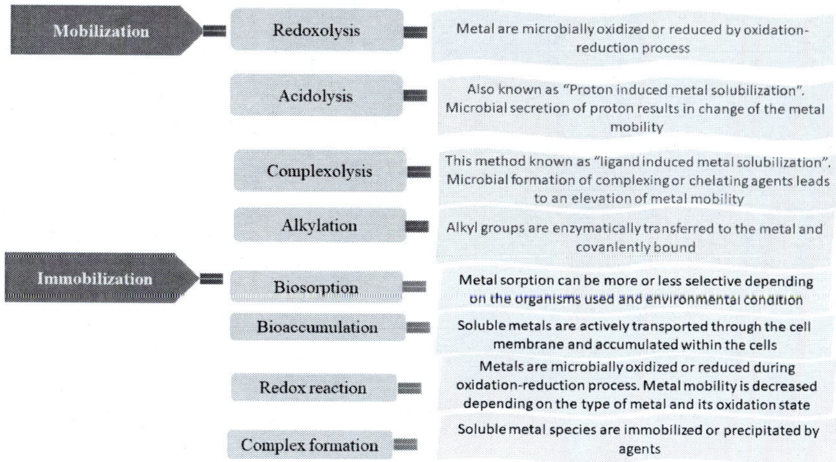

Figure 5. Mobilisation and Immobilisation of metals by microbes.

Bacteria, which naturally form exopolysaccharide coatings, exhibit the potential to bioabsorb metal ions and block them from interrelating with important cellular constituents (Ryan et al., 2015). Das (2010) mentioned that biosorption is an effective technology for cleaning aqueous environments contaminated with metals or even for recovering precious metals. Bacteria have been utilised as bioadsorbents due to their tiny size, the universality of and their propensity to grow under strict conditions, and their resistance to a variety of different environmental conditions (Ryan et al., 2015). The biosorption process, which is the collection of metals that are independent of metabolism, is often fast. In contrast, bioaccumulation is the metabolism-linked intracellular uptake of metal ions by various living microorganisms (Ahemad and Kibret, 2013). Metal-binding proteins allow metal ions to actively migrate within the cell (Herald et al., 2003). Intracellular isolation is the complexation strategy of metals by different compounds in the cytoplasm of cells. Metal concentrations in microbial cells may be due to interaction with surface ligands, followed by slow transport to cells (Gavrilescu, 2004; Ryan et al., 2015).

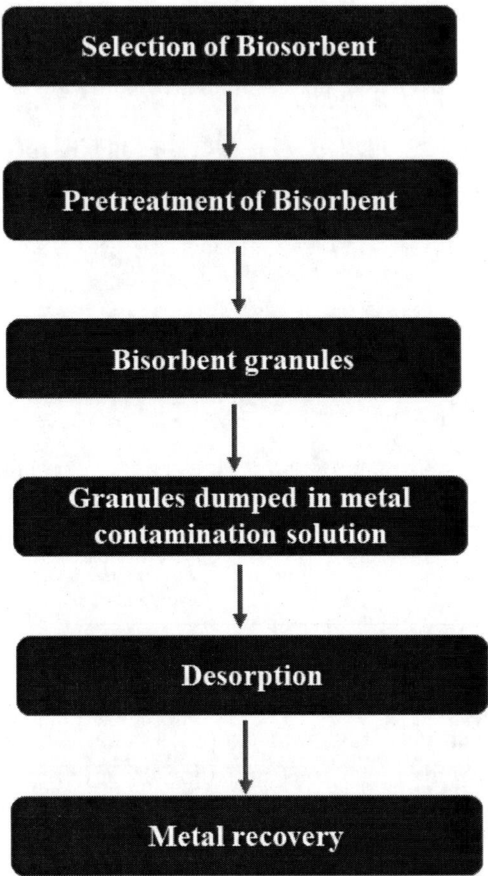

Figure 6. Biosorption process.

In bioleaching, it may be better to use elemental sulphur as a substrate than to treat with sulphuric acid to solubilise metals from polluted aquatic sediments (Gadd, 2000). Bioleaching refers to the transformation of solid metal values to their water-soluble forms with the help of microorganisms. There is growing interest in the use of bioleaching for the biological restoration of solid waste by removing heavy metals from ash and sewage sludge (Gu et al., 2018). The binding of metal to extracellular substances fixes the metal and prevents it from entering the cell. Extracellular binding usually occurs in layers of carbohydrates, polysaccharides, and in some cases mucus or exopolymers composed of nucleic acids and fatty acids (Schiewer and Volesky, 2000). Aforesaid exopolymers, or extracellular polymer substances (EPS), are widely distributed in the natural environment and protect in opposition to microorganisms (Bhaskar and Bhosle, 2005). Methylation promotes the diffusion of metals away from cell, effectively reducing the overall metal toxicity (Kosolapov et al., 2004).

8. Microbial Metal Redox Transformations and Metabolic Availability

Microorganisms are ubiquitous in nature and play an important role in the basic biogeochemical cycle of metal conversion between soluble and insoluble species. Interaction with metal microbes can have positive or detrimental consequences (Issazadeh, et al., 2013). Microorganisms can detoxify metals by valence conversion, extracellular chemical precipitation, or volatilisation. Regarding the presence of minerals and metal compounds in waste, various metabolic process can arise when microorganisms come into proximity with solid or dissolved metals. According to Ayangbenro and Babalola (2017), several metals can be enzymatically reduced during the metabolic processes unrelated to metal assimilation. Oxidised states or organic complexes of heavy metals are never completely destroyed and can be converted to water-soluble, less toxic and precipitated forms (Garbisu and Alkorta, 2001). Distinct types of microbial electron acceptor classes can be engaged in bioremediation, such as oxygen, nitrate, manganese, iron, sulphate, or carbon dioxide-reducing, and their corresponding redox potentials.

In the case of copper, copper sulphide is microbially oxidised into copper sulphate and there are metallic components in the aqueous phase. Biooxidation represents the microbiological oxidation (loss of electrons) of host minerals containing metallic compounds of interest. Like all other cells, microbes can absorb soluble metals for their metabolism. Microbes have two options for overcoming an insufficiency of a peculiar metal with limited solubility in its environment, so they can either stop their growth and wait for the environment to improve or secondly, they can actively enhance the solubility of the metal (Monachese et al., 2012; Kaur, et al., 2020). The resistance mechanisms developed by microorganisms include efflux pumps, permeable barriers, enzymatic detoxification, intracellular and extracellular sequestration and reduction (Ruggaber and Talley, 2006), intracellular and extracellular sequestration and reduction). Essential metals are part of an active enzyme core or are structure-forming elements.

Redox-active metals can take part in an active role in enzymatic reactions, while framework metals are not redox-active. Robinson et al. (2015) stated that the level of essential metals in an organism is carefully modulated because a lack of metal closes off vital metabolic pathways and excess is likely to impede metabolic pathways. Studies of heavy metal accumulation by microorganisms require an expression of phytokeratin and metal-binding proteins and peptides or metallothionein (Cobbett and Goldsbrough, 2002). Hormonal and redox signaling processes are regulated by metallothionein transcriptional factors in connection with exposure to toxic metals (Abdelmigid, 2016). The redox potential of the environment controls the direction of chemical equilibrium, whether the metal is reduced or oxidised. Enzymatic catalytic properties and physiological functions of microorganisms and various environmental factors such as pH, temperature, comet ions, and nutrient sources are key factors in determining heavy metal mobility and bioavailability for microbial conversion (Zhang et al., 2019).

Metal-containing enzymes or metalloenzymes play a crucial role in every metabolic cycle, be it the synthesis of high-energy molecules such as adenosine triphosphate (ATP), deoxyribonucleic acid (DNA), transcription, or protein/carbohydrate synthesis (Glass and Orphan, 2012). Biologically encoded alterations in the state of oxidation help to repair the environmental niches of sediments, soil and water that heavy metals have altered. Electron transfer reactions are fundamental in the respiratory chain in different cells and are independent of respiration. A variety of enzymes catalyses these reactions. Electron transfer is primarily accomplished by cytochrome C family iron-containing enzymes. Ironbound to heme in these enzymes forms the active electron handling, the connecting link between the transfer steps (Karigar and Rao, 2011). Biominerals can result from redox conversion of metals, sorption phenomena, and metabolic activity Metallothionein has a strong affinity for metals such as cadmium, mercury, silver, zinc, and copper. The presence of metals instigates the synthesis of these proteins. Therefore, metal detoxification is their main function. These proteins are detected in *Synechoccus spp., E. coli* and *Pseudomonas putida* (Pepper et al., 2015; Gupta et al., 2016).

When bacterial cells are exhibited to high rise of heavy metals, the intracellular metals react with different metabolites to form toxic compounds.

The diverse mechanism for up taking these metal species is bacterial cell machinery, where heavy metals enter the cell (Issazadeh et al., 2013). Basically, the bioadsorption of heavy metals by the bacterial cell system relies on non-enzymatic processes such as adsorption. Adsorption is distinguished by the non-specific binding of different metal ions to extracellular / cell surface-related polysaccharides and proteins. In a nutshell, biosorption is described as the property of an inert or dead microbial biomass that binds and concentrates heavy metals even from very dilute solutions (Gavrilescu, 2004). Microbial metabolic processes have been extensively studied for metal repairs such as in vivo changes, including microbial oxidation and metal reduction (Ahemad and Kibret, 2013). The adsorbed heavy metal ions are metabolised to living bacterial cells and can change the redox state of the heavy metal ions to reduce toxicity.

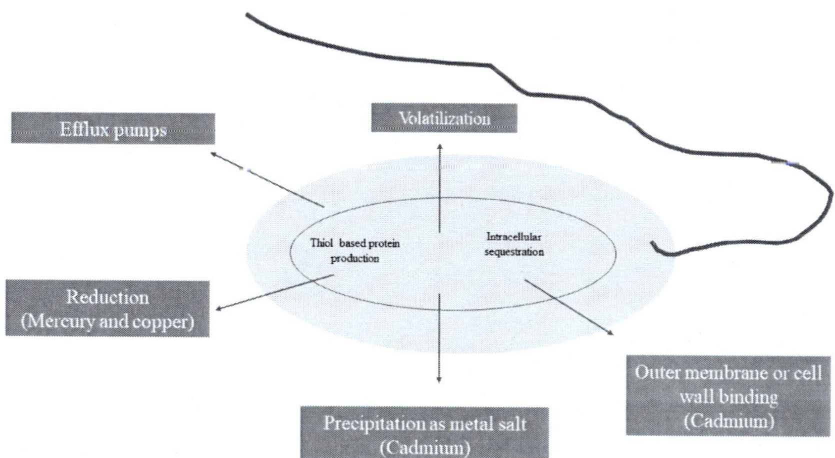

Figure 7. Mechanisms (intracellular and extracellular) developed by microorganism to resist and detoxify harmful heavy metal.

Microorganisms can initiate mobilisation/immobilisation of metal via the redox reaction. Therefore, it affects the bioremediation process. Heavy metals such as Fe, As, Cr, and Hg go through an oxidation and reduction cycle (Bolan et al., 2014). The bioavailability of metals changes in response to changing redox conditions. Under oxidative or aerobic conditions, many metals are often found as soluble cationic forms (De Jonge et al., 2012). Reduction of metals, including As, Cr, Hg, and Se, are most often exposed to microbial oxidation/reduction reactions, affecting their speciation and mobility.

9. Genetics of Metal Resistance or Eco Genomics of Microbes

The microbial ecogenomics group combines interdisciplinary research disciplines to study modern and ancient microbial communities. A microbial ecosystem can be defined as a system of all microorganisms that live in a particular area or niche and interact in the context of other organisms and abiotic. The metal resistance of microorganisms is heterogeneous in both the genetic and biochemical bases of those encoded by chromosomes, plasmids, or transposons, and may involve one or more genes. Microbial ecogenomics leverages on the recent advances in high-throughput and output genome technology combined with microbial physiology research to address these complex bioremediation issues at the system level (Mende et al., 2016). Cloning and sequencing approaches are applied to identify or understand large phylogenetic clusters of organisms responsible for various metal oxidation/reduction mechanisms in the presence of metal.

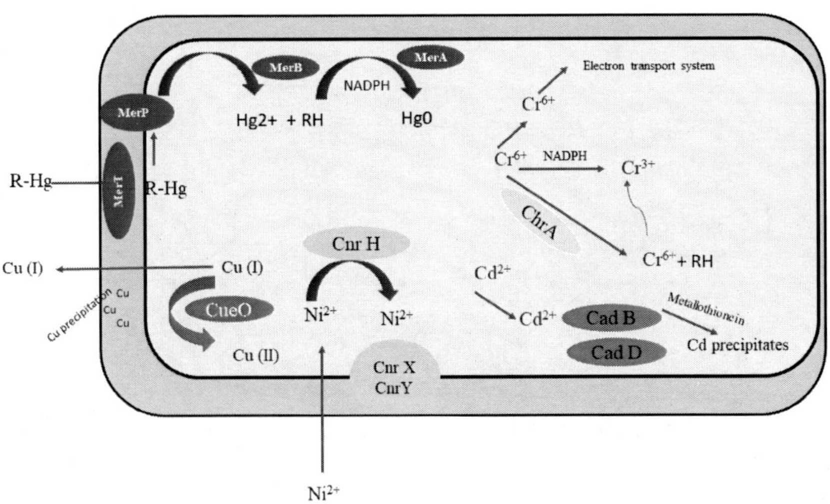

Figure 8. Genetic mechanism of resistance to toxic metals by bacteria.

The metagenomics application provides a comprehensive overview of the genetic makeup of the microbial community, including information on the identities and potential metabolic capacity of community members (Zhou et al., 2015). The metagenomics approach has been used to study microbial communities associated with different environments. Ecogenomics seeks the definition of the abundance and identity of the microbial flora in environmental samples. There are two technical approaches to start an ecogenomics technique. Polymerase chain reaction (PCR) sequence and shotgun sequence (Kumar et al., 2016). Some bacteria species can actively transport toxic metal to the outside of the cells due to ATPase outflow mechanism and ion pump expression. Genes for such biological exclusion mechanisms are commonly encoded in plasmids (Kushwaha et al., 2017).

10. Genetic Engineering in Metal–Microbe Interaction

Researchers now have the opportunity to grow microorganisms that are significant for bioremediation and can assess their physiology using a mixture of genome-based modeling and experimental procedures. Genetic engineering integrated into bioremediation supports the manipulation of the bacterial genome. This can improve the detoxification of toxic metals, which is not usually done with normal bacteria (Das et al., 2016). In addition, new environmental genomic techniques offer the possibility of carrying out similar studies in organisms not yet cultured. Bioremediation technology uses microorganisms to reduce or eliminate harmless products such as halogenated contaminants that are present in soil, sediment, water or air, or to convert them into harmless products (Sharma, 2012). Traditional methods of heavy metal repair have many limitations, including the production of toxic chemical sludge, and are not environmentally friendly (Tchounwou et al., 2012; Ojuederie and Babalola, 2017). *Bacillus subtilis* has been genetically engineered to produce the thermostable enzyme methyltransferase.

In 1991, metabolic engineering, first proposed by James E. Bailey as a new scientific discipline for improving genetic and regulatory processes within the cells pointed two ways (a) Enhance the yield and productivity of native products integrated by organisms, (b) establish the synthesis of products new to the host cell (Bailey, 1996). Moreover, now researchers have also manipulated microorganisms to have chimeric or recombinant metal-binding proteins and peptide linkages on the extracellular surface, thereby revamping the ability and particularity of these microbial bio adsorbents (Ueda, 2016).

The recombinant expression of target genes in transgenic microorganisms allows for improved uptake and sequestration of heavy metal ions. Biologically powered heavy metal removal technology is often said to be inexpensive, environmentally friendly and easy to use, using biomass to remove heavy metal contaminants from wastewater. Efforts to improve the uptake of biological heavy metals have been made from periplasm to gram using recombinant expression intimal importers from three major classes: transporters, channels, secondary carriers, and primary active transporters. The focus has been on improving the uptake of negative bacteria into the cytoplasm (Saier, 2016).

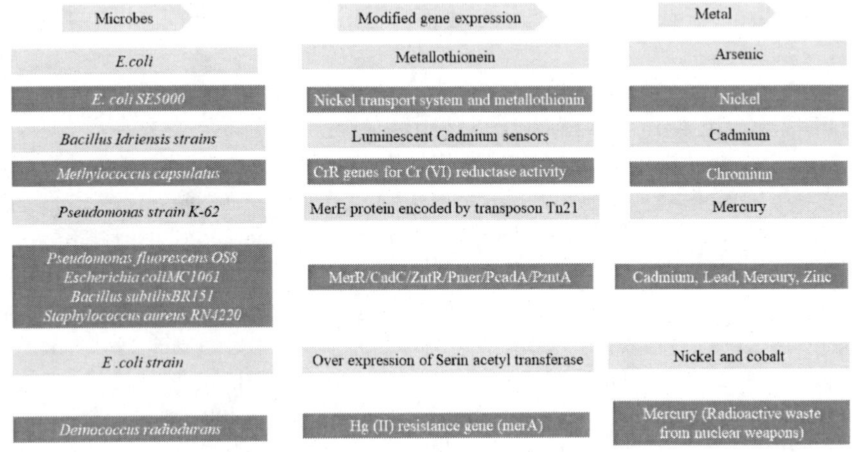

Figure 9. Genetically modified organisms used in metal bioremediation procedure.

Modifications in microbial composition occur fundamentally throughout blocking essential functional groups, the substitution of essential metal ions, or exerting inhibitory effects by altering the active conformation of biomolecules. Bae and his team (2001) reported the impact of genetically modified *Escherichia coli* in uptake and bioaccumulation of mercury. The gene encoding cytochrome c7 from *Desulphuromonas acetooxidans* has been cloned and expressed in *Desulphovibrio desulphuricans,* the recombinant organism showed magnified expression of metal reductase activities (Lloyd et al., 2003).

11. Impact of Metal–Microbe Interaction on Ecosystem: Both Positive and Negative

Metabolism and growth can lead to alterations in pH level, redox potential and ionic strength of the soil system. Bacteria, both aerobic and anaerobic, can use sulphur, iron, fatty acids, or hydrogen as a carbon source. The acetic acid oxidation step of microorganisms is crucial in the complete detoxification of acidic waste in the mine, producing bicarbonate ions that neutralise the acidic waste. Bioremediation is used to transform toxic heavy metals into less harmful states by purifying the polluted environment using the microorganism or its enzymes. Bioremediation via biofilm can be used to purify an environment contaminated with heavy metals.

Diverse microbial species within the biofilm consortium make different key roles, making biofilms robust and resistant to harmful environmental factors. The mechanism of microbial immobilisation and solubilisation of metals, radionuclides, and associated substances has clear potential for bioremediation. Several procedures are an essential part of the operation of some successful in-situ and ex-situ procedures. Biodegradation is environmentally friendly and cheap to revitalise the environment (Begum et al., 2021). Soil microbial complexes are crucial for agriculture, waste management and water management. Environmental biotechnology uses these microbial activities in a targeted manner to process heavy metals. Moreover, the utilisation of microbial fuel cells to break down or chelate intractable heavy metals has been established.

Today, microbial technology is active and growing. The use of microorganisms as a green or eco-based approach for the synthesis of metal nanoparticles has been described (Klaus-Joerger et al., 2001). Bioremediation is very effective because it depends only on microorganisms naturally present in the soil and does not pose a threat to the environment or the people living in the area, as environmental conditions are only effective if they allow growth and activity of the microorganisms. All microbial-based bioremediation technologies of metals have their own strengths and weaknesses because they have their own specific uses.

Conclusion

Microorganisms play a crucial role in global activity in the biosphere, especially in biogeochemical cycles of elements, metal-mineral conversion, and degradation. On a larger scale, much research has been done to improve the understanding of the nature of microorganisms as they interact with various toxic pollutants. On the other hand, it is necessary to understand the mechanism of microbial response to heavy metal exposure and encourage research to validate the screening of resistant microorganisms that may be used for both restoration and remediation.

References

Abdelmigid, Hala M. "Expression analysis of type 1 and 2 metallothionein genes in rapeseed (Brassica napus L.) during short-term stress using sqRT-PCR analysis." *Indian J Exp Biol.* (2016); 54(3):212-8.

Ahemad, Munees, and Mulugeta Kibret. "Recent trends in microbial biosorption of heavy metals: a review." *Biochemistry and Molecular Biology* 1, no. 1 (2013): 19-26.

Ahemad, Munees, and Mulugeta Kibret. "Recent trends in microbial biosorption of heavy metals: a review." *Biochemistry and Molecular Biology* 1, no. 1 (2013): 19-26.

Ayangbenro, Ayansina Segun, and Olubukola Oluranti Babalola. "A new strategy for heavy metal polluted environments: a review of microbial biosorbents." *International journal of environmental research and public health* 14, no. 1 (2017): 94.

Bae, Weon, Rajesh K. Mehra, Ashok Mulchandani, and Wilfred Chen. "Genetic engineering of Escherichia coli for enhanced uptake and bioaccumulation of mercury." *Applied and environmental microbiology* 67, no. 11 (2001): 5335-5338.

Bailey, James E. "Metabolic engineering." In *Advances in Molecular and Cell Biology*, vol. 15, pp. 289-296. Elsevier, 1996.

Becerra-Castro, C., Monterroso, C., Garcia-Leston, M., Prieto-Fernandez, A., Acea, M. J., & Kidd, P. S. (2009). Rhizosphere microbial densities and trace metal tolerance of the nickel hyperaccumulator Alyssum serpyllifolium subsp. lusitanicum. *International journal of phytoremediation*, *11*(6), 525-541.

Begum, Shahani, Sakti Kanta Rath, and Chandi Charan Rath. "Applications of Microbial Communities for the Remediation of Industrial and Mining Toxic Metal Waste: A Review." *Geomicrobiology Journal* (2021): 1-12.

Bento, Fatima Menezes, Flávio Anastácio de Oliveira Camargo, Benedict Okeke, and Willian Thomas Frankenberger-Júnior. "Bioremediation of soil contaminated by diesel oil." *Brazilian journal of Microbiology* 34 (2003): 65-68.

Bhaskar, P. V., and Narayan B. Bhosle. "Microbial extracellular polymeric substances in marine biogeochemical processes." *Current Science* (2005): 45-53.

Bolan, Nanthi, Anitha Kunhikrishnan, Ramya Thangarajan, Jurate Kumpiene, Jinhee Park, Tomoyuki Makino, Mary Beth Kirkham, and Kirk Scheckel. "Remediation of heavy metal (loid) s contaminated soils–to mobilize or to immobilize?." *Journal of hazardous materials* 266 (2014): 141-166.

Boteva, Silvena, Galina Radeva, Ivan Traykov, and Anelia Kenarova. "Effects of long-term radionuclide and heavy metal contamination on the activity of microbial communities, inhabiting uranium mining impacted soils." *Environmental Science and Pollution Research* 23, no. 6 (2016): 5644-5653.

Bottjer, David J. "Geobiology and the fossil record: eukaryotes, microbes, and their interactions." *Palaeogeography, Palaeoclimatology, Palaeoecology* 219, no. 1-2 (2005): 5-21.

Brandl, H. (2001). Microbial leaching of metals. *Biotechnology*, *10*, 191-224.

Chellaiah, Edward Raja. "Cadmium (heavy metals) bioremediation by Pseudomonas aeruginosa: a minireview." *Applied water science* 8, no. 6 (2018): 1-10.

Chen, Shen-Yi, and Jih-Gaw Lin. "Bioleaching of heavy metals from contaminated sediment by indigenous sulfur-oxidizing bacteria in an air-lift bioreactor: effects of sulfur concentration." *Water Research* 38, no. 14-15 (2004): 3205-3214.

Cheung, K. H., and Ji-Dong Gu. "Mechanism of hexavalent chromium detoxification by microorganisms and bioremediation application potential: a review." *International Biodeterioration & Biodegradation* 59, no. 1 (2007): 8-15.

Chorover, Jon, Ruben Kretzschmar, Ferran Garcia-Pichel, and Donald L. Sparks. "Soil biogeochemical processes within the critical zone." *Elements* 3, no. 5 (2007): 321-326.

Cobbett, Christopher, and Peter Goldsbrough. "Phytochelatins and metallothioneins: roles in heavy metal detoxification and homeostasis." *Annual review of plant biology* 53, no. 1 (2002): 159-182.

Cyriaque, Valentine, Augustin Géron, Gabriel Billon, Joseph Nesme, Johannes Werner, David C. Gillan, Søren J. Sørensen, and Ruddy Wattiez. "Metal-induced bacterial interactions promote diversity in river-sediment microbiomes." *FEMS microbiology ecology* 96, no. 6 (2020): fiaa076.

Das, N. (2010). Recovery of precious metals through biosorption—a review. *Hydrometallurgy*, *103*(1-4), 180-189.

Das, Surajit, Hirak R. Dash, and Jaya Chakraborty. "Genetic basis and importance of metal resistant genes in bacteria for bioremediation of contaminated environments with toxic metal pollutants." *Applied microbiology and biotechnology* 100, no. 7 (2016): 2967-2984.

De Jonge, M., J. Teuchies, P. Meire, R. Blust, and L. Bervoets. "The impact of increased oxygen conditions on metal-contaminated sediments part I: effects on redox status, sediment geochemistry and metal bioavailability." *Water research* 46, no. 7 (2012): 2205-2214.

Fabietti, G., Biasioli, M., Barberis, R., & Ajmone-Marsan, F. (2010). Soil contamination by organic and inorganic pollutants at the regional scale: the case of Piedmont, Italy. *Journal of Soils and Sediments,* 10(2), 290-300.

Ferris, F. G., and T. J. Beveridge. "Site specificity of metallic ion binding in Escherichia coli K-12 lipopolysaccharide." *Canadian journal of microbiology* 32, no. 1 (1986): 52-55.

Gadd, Geoffrey M., and Jacqueline A. Sayer. "Influence of fungi on the environmental mobility of metals and metalloids." *Environmental Microbe-Metal Interactions* (2000): 237-256.

Gadd, Geoffrey Michael, and Xiangliang Pan. "Biomineralization, bioremediation and biorecovery of toxic metals and radionuclides." *Geomicrobiology Journal* (2016): 175-178.

Gadd, Geoffrey Michael. "Bioremedial potential of microbial mechanisms of metal mobilization and immobilization." *Current opinion in biotechnology* 11, no. 3 (2000): 271-279.

Gadd, Geoffrey Michael. "Metals, minerals and microbes: geomicrobiology and bioremediation." *Microbiology* 156, no. 3 (2010): 609-643.

Gadd, Geoffrey Michael. "Microbial roles in mineral transformations and metal cycling in the Earth's critical zone." In *Molecular environmental soil science*, pp. 115-165. Springer, Dordrecht, 2013.

Garbisu, Carlos, and Itziar Alkorta. "Phytoextraction: a cost-effective plant-based technology for the removal of metals from the environment." *Bioresource technology* 77, no. 3 (2001): 229-236.

Gavrilescu, Maria. "Removal of heavy metals from the environment by biosorption." *Engineering in Life Sciences* 4, no. 3 (2004): 219-232.

Glass, J., & Orphan, V. J. (2012). Trace metal requirements for microbial enzymes involved in the production and consumption of methane and nitrous oxide. *Frontiers in microbiology*, *3*, 61.

Gomathy, M., and K. G. Sabarinathan. "Microbial mechanisms of heavy metal tolerance-a review." *Agricultural Reviews* 31, no. 2 (2010): 133-138.

Griffitt, Robert J., Jing Luo, Jie Gao, Jean-Claude Bonzongo, and David S. Barber. "Effects of particle composition and species on toxicity of metallic nanomaterials in aquatic organisms." *Environmental Toxicology and Chemistry: An International Journal* 27, no. 9 (2008): 1972-1978.

Group, The MerMex, X. Durrieu de Madron, C. Guieu, R. Sempéré, P. Conan, D. Cossa, F. D'Ortenzio et al. "Marine ecosystems' responses to climatic and anthropogenic forcings in the Mediterranean." *Progress in Oceanography* 91, no. 2 (2011): 97-166.

Gu, Tingyue, Seyed Omid Rastegar, Seyyed Mohammad Mousavi, Ming Li, and Minghua Zhou. "Advances in bioleaching for recovery of metals and bioremediation of fuel ash and sewage sludge." *Bioresource technology* 261 (2018): 428-440.

Gupta, Abhijit, Jyoti Joia, Aditya Sood, Ridhi Sood, Candy Sidhu, and Gaganjot Kaur. "Microbes as potential tool for remediation of heavy metals: a review." *J Microb Biochem Technol* 8, no. 4 (2016): 364-372.

Haferburg, G., & Kothe, E. (2007). Microbes and metals: interactions in the environment. *Journal of basic microbiology*, *47*(6), 453-467.

Hall, J. Á. (2002). Cellular mechanisms for heavy metal detoxification and tolerance. *Journal of experimental botany*, *53*(366), 1-11.

He, Z. L., Yang, X. E., & Stoffella, P. J. (2005). Trace elements in agroecosystems and impacts on the environment. *Journal of Trace elements in Medicine and Biology*, *19*(2-3), 125-140.

Hemambika, B., M. Johncy Rani, and V. Rajesh Kannan. "Biosorption of heavy metals by immobilized and dead fungal cells: A comparative assessment." *Journal of Ecology and the Natural Environment* 3, no. 5 (2011): 168-175.

Herald, V. L., J. L. Heazlewood, D. A. Day, and A. H. Millar. "Proteomic identification of divalent metal cation binding proteins in plant mitochondria." *FEBS letters* 537, no. 1-3 (2003): 96-100.

Hinojosa, M. Belén, José A. Carreira, Roberto García-Ruíz, and Richard P. Dick. "Soil moisture pre-treatment effects on enzyme activities as indicators of heavy metal-contaminated and reclaimed soils." *Soil Biology and Biochemistry* 36, no. 10 (2004): 1559-1568.

Hoehler, T. M., & Jørgensen, B. B. (2013). Microbial life under extreme energy limitation. *Nature Reviews Microbiology*, *11*(2), 83-94.

Holtan-Hartwig, L., Bechmann, M., Høyås, T. R., Linjordet, R., & Bakken, L. R. (2002). Heavy metals tolerance of soil denitrifying communities: N2O dynamics. *Soil Biology and Biochemistry*, *34*(8), 1181-1190.

Hoostal, M. J., Bidart-Bouzat, M. G., & Bouzat, J. L. (2008). Local adaptation of microbial communities to heavy metal stress in polluted sediments of Lake Erie. *FEMS microbiology ecology*, *65*(1), 156-168.

Issazadeh, Khosro, Nadiya Jahanpour, Fataneh Pourghorbanali, Golnaz Raeisi, and Jamileh Faekhondeh. "Heavy metals resistance by bacterial strains." *Annals of Biological Research* 4, no. 2 (2013): 60-63.

Jaiswal, Deepa, and Jitendra Pandey. "Impact of heavy metal on activity of some microbial enzymes in the riverbed sediments: Ecotoxicological implications in the Ganga River (India)." *Ecotoxicology and environmental safety* 150 (2018): 104-115.

Jasrotia, Shivakshi, Arun Kansal, and V. V. N. Kishore. "Arsenic phyco-remediation by Cladophora algae and measurement of arsenic speciation and location of active absorption site using electron microscopy." *Microchemical journal* 114 (2014): 197-202.

Karigar, C. S., & Rao, S. S. (2011). Role of microbial enzymes in the bioremediation of pollutants: a review. *Enzyme research, 2011.*

Kaur, T., Rana, K. L., Kour, D., Sheikh, I., Yadav, N., Kumar, V., ... & Saxena, A. K. (2020). Microbe-mediated biofortification for micronutrients: present status and future challenges. In *New and Future Developments in Microbial Biotechnology and Bioengineering* (pp. 1-17). Elsevier.

Ke, Xin, Shaofeng Gui, Hao Huang, Haijun Zhang, Chunyong Wang, and Wei Guo. "Ecological risk assessment and source identification for heavy metals in surface sediment from the Liaohe River protected area, China." *Chemosphere* 175 (2017): 473-481.

Kido, S. (2013). Secondary osteoporosis or secondary contributors to bone loss in fracture. Bone metabolism and heavy metals (cadmium and iron). *Clinical calcium*, *23*(9), 1299-1306.

Konhauser, K. "Cell surface reactivity and metal sorption." *Introduction to Geomicrobiology.* Blackwell: Oxford (2007).

Kosolapov, D. B., P. Kuschk, M. B. Vainshtein, A. V. Vatsourina, A. Wiessner, M. Kästner, and R. A. Müller. "Microbial processes of heavy metal removal from carbon-deficient effluents in constructed wetlands." *Engineering in Life Sciences* 4, no. 5 (2004): 403-411.

Kumar, Manoj, Ajit Varma, and Vivek Kumar. "Ecogenomics based microbial enzyme for biofuel industry." *Sci Int* 4 (2016): 1-11.

Kushwaha, Anamika, Radha Rani, Sanjay Kumar, Tarence Thomas, Arun Alfred David, and Meraz Ahmed. "A new insight to adsorption and accumulation of high lead concentration by exopolymer and whole cells of lead-resistant bacterium Acinetobacter junii L. Pb1 isolated from coal mine dump." *Environmental Science and Pollution Research* 24, no. 11 (2017): 10652-10661.

laus-Joerger, Tanja, Ralph Joerger, Eva Olsson, and Claes Göran Granqvist. "Bacteria as workers in the living factory: metal-accumulating bacteria and their potential for materials science." *TRENDS in Biotechnology* 19, no. 1 (2001): 15-20.

Li, Kefeng, and Wusirika Ramakrishna. "Effect of multiple metal resistant bacteria from contaminated lake sediments on metal accumulation and plant growth." *Journal of hazardous materials* 189, no. 1-2 (2011): 531-539.

Lloyd, Jon R., Ching Leang, Allison L. Hodges Myerson, Maddalena V. Coppi, Stacey Cuifo, Barb Methe, Steven J. Sandler, and Derek R. Lovley. "Biochemical and genetic characterization of PpcA, a periplasmic c-type cytochrome in *Geobacter sulfurreducens*." *Biochemical Journal* 369, no. 1 (2003): 153-161.

Lovley, D. R. (2003). Cleaning up with genomics: applying molecular biology to bioremediation. *Nature Reviews Microbiology*, *1*(1), 35-44.

Maanan, M., Landesman, C., Maanan, M., Zourarah, B., Fattal, P., & Sahabi, M. (2013). Evaluation of the anthropogenic influx of metal and metalloid contaminants into the Moulay Bousselham lagoon, Morocco, using chemometric methods coupled to geographical information systems. *Environmental Science and Pollution Research*, *20*(7), 4729-4741.

Maodaa, Saleh N., Ahmed A. Allam, Jamaan Ajarem, Mostafa A. Abdel-Maksoud, Gadah I. Al-Basher, and Zun Yao Wang. "Effect of parsley (Petroselinum crispum, Apiaceae) juice against cadmium neurotoxicity in albino mice (Mus musculus)." *Behavioral and Brain Functions* 12, no. 1 (2016): 1-16.

Mende, Daniel R., Frank O. Aylward, John M. Eppley, Torben N. Nielsen, and Edward F. DeLong. "Improved environmental genomes via integration of metagenomic and single-cell assemblies." *Frontiers in microbiology* 7 (2016): 143.

Mishra, S., Bharagava, R. N., More, N., Yadav, A., Zainith, S., Mani, S., & Chowdhary, P. (2019). Heavy metal contamination: an alarming threat to environment and human health. In *Environmental biotechnology: For sustainable future* (pp. 103-125). Springer, Singapore.

Monachese, M., Burton, J. P., & Reid, G. (2012). Bioremediation and tolerance of humans to heavy metals through microbial processes: a potential role for probiotics?. *Applied and environmental microbiology*, *78*(18), 6397-6404.

Morel, François M. M., and N. M. Price. The biogeochemical cycles of trace metals in the oceans. *Science* 300, no. 5621 (2003): 944-947.

Moreno-Mateos, David, Edward B. Barbier, Peter C. Jones, Holly P. Jones, James Aronson, José A. López-López, Michelle L. McCrackin, Paula Meli, Daniel Montoya, and José M. Rey Benayas. "Anthropogenic ecosystem disturbance and the recovery debt." *Nature communications* 8, no. 1 (2017): 1-6.

Nascimento, Andréa M. A., and Edmar Chartone-Souza. "Operon mer: bacterial resistance to mercury and potential for bioremediation of contaminated environments." *Genetics and molecular research* 2, no. 1 (2003): 92-101.

Nejad, Zahra Derakhshan, Myung Chae Jung, and Ki-Hyun Kim. "Remediation of soils contaminated with heavy metals with an emphasis on immobilization technology." *Environmental geochemistry and health* 40, no. 3 (2018): 927-953.

Ojuederie, Omena Bernard, and Olubukola Oluranti Babalola. "Microbial and plant-assisted bioremediation of heavy metal polluted environments: a review." *International journal of environmental research and public health* 14, no. 12 (2017): 1504.

Orell, Alvaro, Claudio A. Navarro, Rafaela Arancibia, Juan C. Mobarec, and Carlos A. Jerez. "Life in blue: copper resistance mechanisms of bacteria and archaea used in industrial biomining of minerals." *Biotechnology advances* 28, no. 6 (2010): 839-848.

Osman, Gamal E. H., Hussein H. Abulreesh, Khaled Elbanna, and Mohammed R. Shaaban. "Samreen, Iqbal Ahmad, Recent Progress in Metal-Microbe Interactions: Prospects in Bioremediation." *J Pure Appl Microbiol* 13, no. 1 (2019): 13-26.

Pepper, Ian L., and Terry J. Gentry. "Earth environments." In *Environmental Microbiology*, pp. 59-88. Academic Press, 2015.

Pushkar, Bhupendra, Pooja Sevak, and Akansha Singh. "Bioremediation treatment process through mercury-resistant bacteria isolated from Mithi river." *Applied Water Science* 9, no. 4 (2019): 1-10.

Raab, A., & Feldmann, J. (2003). Microbial transformation of metals and metalloids. *Science progress*, *86*(3), 179-202.

Rahman, Zeeshanur, and Ved Pal Singh. "The relative impact of toxic heavy metals (THMs)(arsenic (As), cadmium (Cd), chromium (Cr)(VI), mercury (Hg), and lead (Pb)) on the total environment: an overview." *Environmental monitoring and assessment* 191, no. 7 (2019): 1-21.

Reddy, M. Sudhakara, Varenyam Achal, and Abhijit Mukherjee. "Microbial concrete, a wonder metabolic product that remediates the defects in building structures." In *Microorganisms in environmental management*, pp. 547-568. Springer, Dordrecht, 2012.

Roane, T. M., Pepper, I. L., & Gentry, T. J. (2015). Microorganisms and metal pollutants. In *Environmental microbiology* (pp. 415-439). Academic Press.

Roane, Timberley M., Ian L. Pepper, and Terry J. Gentry. "Microorganisms and metal pollutants." In *Environmental microbiology*, pp. 415-439. Academic Press, 2015.

Robinson, B. H., Bañuelos, G., Conesa, H. M., Evangelou, M. W., & Schulin, R. (2009). The phytomanagement of trace elements in soil. *Critical Reviews in Plant Sciences*, *28*(4), 240-266.

Robinson, Brett H. "E-waste: an assessment of global production and environmental impacts." *Science of the total environment* 408, no. 2 (2009): 183-191.

Ruggaber, Timothy P., and Jeffrey W. Talley. "Enhancing bioremediation with enzymatic processes: a review." *Practice Periodical of Hazardous, Toxic, and Radioactive Waste Management* 10, no. 2 (2006): 73-85.

Ryan, P. M., R. P. Ross, G. F. Fitzgerald, N. M. Caplice, and C. Stanton. "Sugar-coated: exopolysaccharide producing lactic acid bacteria for food and human health applications." *Food & function* 6, no. 3 (2015): 679-693.

Saier Jr, Milton H. "Transport protein evolution deduced from analysis of sequence, topology and structure." *Current opinion in structural biology* 38 (2016): 9-17.

Schiewer, Silke, and Bohumil Volesky. "Biosorption processes for heavy metal removal." *Environmental microbe-metal interactions* (2000): 329-362.

Sharma, S. (2012). Bioremediation: features, strategies and applications. *Asian Journal of Pharmacy and Life Science ISSN*, *2231*, 4423.

Singh, Reena, Neetu Gautam, Anurag Mishra, and Rajiv Gupta. "Heavy metals and living systems: An overview." *Indian journal of pharmacology* 43, no. 3 (2011): 246.

Sprocati, A. R., Alisi, C., Segre, L., Tasso, F., Galletti, M., & Cremisini, C. (2006). Investigating heavy metal resistance, bioaccumulation and metabolic profile of a metallophile microbial consortium native to an abandoned mine. *Science of the total environment*, *366*(2-3), 649-658.

Sugio, Tsuyoshi, Taher M. Taha, Atsunori Negishi, and Fumiaki Takeuchi. "Existence of ferrous iron-dependent mercury reducing enzyme system in sulfur-grown A. ferrooxidans MON-1 cells." In *Advanced Materials Research*, vol. 71, pp. 745-748. Trans Tech Publications Ltd, 2009.

Tamás, Markus J., Sandeep K. Sharma, Sebastian Ibstedt, Therese Jacobson, and Philipp Christen. "Heavy metals and metalloids as a cause for protein misfolding and aggregation." *Biomolecules* 4, no. 1 (2014): 252-267.

Tan, Yang, Ramesh Y. Adhikari, Nikhil S. Malvankar, Joy E. Ward, Kelly P. Nevin, Trevor L. Woodard, Jessica A. Smith et al. "The low conductivity of Geobacter uraniireducens pili suggests a diversity of extracellular electron transfer mechanisms in the genus Geobacter." *Frontiers in microbiology* 7 (2016): 980.

Tchounwou, Paul B., Clement G. Yedjou, Anita K. Patlolla, and Dwayne J. Sutton. "Heavy metal toxicity and the environment." *Molecular, clinical and environmental toxicology* (2012): 133-164.

Tobor-Kapłon, Maria A., Jaap Bloem, Paul F. A. M. Römkens, and P. C. de Ruiter. "Functional stability of microbial communities in contaminated soils." *Oikos* 111, no. 1 (2005): 119-129.

Turpeinen, Riina, Timo Kairesalo, and Max M. Häggblom. "Microbial community structure and activity in arsenic-, chromium-and copper-contaminated soils." *FEMS Microbiology Ecology* 47, no. 1 (2004): 39-50.

Ueda, Mitsuyoshi. "Establishment of cell surface engineering and its development." *Bioscience, biotechnology, and biochemistry* 80, no. 7 (2016): 1243-1253.

Vickers, Neil J. "Animal communication: when i'm calling you, will you answer too?." *Current biology* 27, no. 14 (2017): R713-R715.

Wang, Shuai-Long, Xiang-Rong Xu, Yu-Xin Sun, Jin-Ling Liu, and Hua-Bin Li. "Heavy metal pollution in coastal areas of South China: a review." *Marine pollution bulletin* 76, no. 1-2 (2013): 7-15.

Wang, Yong. "Optimization of cadmium, zinc and copper biosorption in an aqueous solution by Saccharomyces cerevisiae." *International journal of chemistry* 1 (2012): 1-13.

Warren, Lesley A., and Elizabeth A. Haack. "Biogeochemical controls on metal behaviour in freshwater environments." *Earth-Science Reviews* 54, no. 4 (2001): 261-320.

Wei, YiHua, JinYan Zhang, DaWen Zhang, TianHua Tu, and LinGuang Luo. "Metal concentrations in various fish organs of different fish species from Poyang Lake, China." *Ecotoxicology and environmental safety* 104 (2014): 182-188.

Wichard, Thomas. "Exploring bacteria-induced growth and morphogenesis in the green macroalga order Ulvales (Chlorophyta)." *Frontiers in plant science* 6 (2015): 86.

Wong, C. S., Li, X., & Thornton, I. (2006). Urban environmental geochemistry of trace metals. *Environmental pollution, 142*(1), 1-16.

Xu, Yilu, Balaji Seshadri, Nanthi Bolan, Binoy Sarkar, Yong Sik Ok, Wei Zhang, Cornelia Rumpel et al. "Microbial functional diversity and carbon use feedback in soils as affected by heavy metals." *Environment international* 125 (2019): 478-488.

Yang, Ting, Ming-Li Chen, and Jian-Hua Wang. "Genetic and chemical modification of cells for selective separation and analysis of heavy metals of biological or environmental significance." *TrAC Trends in Analytical Chemistry* 66 (2015): 90-102.

Yin, Kun, Qiaoning Wang, Min Lv, and Lingxin Chen. "Microorganism remediation strategies towards heavy metals." *Chemical Engineering Journal* 360 (2019): 1553-1563.

Zhang, Lixiao, Xianwei Wang, Ramón Cueto, Comfort Effi, Yuling Zhang, Hongmei Tan, Xuebin Qin, Yong Ji, Xiaofeng Yang, and Hong Wang. "Biochemical basis and metabolic interplay of redox regulation." *Redox biology* 26 (2019): 101284.

Zhou, Jizhong, Zhili He, Yunfeng Yang, Ye Deng, Susannah G. Tringe, and Lisa Alvarez-Cohen. "High-throughput metagenomic technologies for complex microbial community analysis: open and closed formats." *MBio* 6, no. 1 (2015): e02288-14.

Chapter 3

Heavy Metal Pollution and Public Health: A Review of Heavy Metal Pollution, Health Implications, and Methods Potentially Used for Pollution Assessment

Innocent Mugudamani[1,*], Saheed. A. Oke[2,†] and Thandi. P. Gumede[1]

[1]Department of Life Sciences,
Central University of Technology, Free State, South Africa
[2]Department of Civil Engineering,
Centre for Sustainable Smart Cities,
Central University of Technology, Free State, South Africa

Abstract

With the rapid industrialization and economic development, heavy metals are continuing to be introduced to soils and sediments through fertilization, irrigation, rivers, runoff, atmospheric deposition and point sources. Additionally, activities such as metal mining, refining, and refinishing by products also contribute to the introduction of heavy metals in the environment. All these activities decrease the capability of the environment to support life thus threatening people, animal and plant health. Health implications associated with heavy metal exposure include those that disturbs nervous, blood forming, cardiovascular, renal and reproductive systems. Furthermore, accumulation of heavy metals in soil diminish quality of soil, cause crop yield decrease and affect the quality

* Corresponding Author's Email: imugudamani@gmail.com; TGumede@cut.ac.za.
† Corresponding Author's Email: soke@cut.ac.za; okesaheed@gmail.com.

In: Trace Metals: Sources, Applications and Environmental Implications
Editor: Oscar M. Thygesen
ISBN: 978-1-68507-797-6

of agricultural products. It is important to appraise the concentration of heavy metals in the environment as they are toxic, persistent and non-degradable. Pollution indices are effective in appraisal of soil pollution with heavy metals, monitoring quality of soil and ensuring future sustainability. This chapter seeks to address the pollution of heavy metals, emission source, health implications and commonly used methods for assessment of pollution and for health risks assessment.

Keywords: heavy metal pollution, public health, anthropogenic, health implications, pollution index, enrichment factor

Introduction

Heavy Metal Pollution and Public Health

The area where living (people, animals and plants) and non-living organisms (soil, water and air) live is referred to as the environment. The environment is clearly distinguished by the biosphere, atmosphere, lithosphere and hydrosphere. The biosphere is considered to be important as it is the area where living organisms intermingle with each other and the non-living organisms (Masindi and Muedi, 2018). Industrialisation and globalisation have damaged environment and their capability to nurture life. Moreover, they have presented environmental contaminants that disturb the overall operation of environment (Sands, 2003; Masindi and Muedi, 2018). Environmental contaminants are chemicals that are available at high concentration than in any fragment of the environment (Martin and Johnson, 2012; Masindi and Muedi, 2018; Briffa, et al., 2020).

The world has experienced an increasing ecological and global public health associated with environmental contamination by heavy metals (Tchounwou et al., 2014). Concerns about the accumulation of heavy metals in soils are due to its persistence and potential toxicity (Ferreira Baptista and De Miguel, 2005; Gabarron, et al., 2017). Heavy metals in the environment occur as a result of natural activities, mining activities, industrial activities, agricultural activities, domestic effluents, pharmaceutical and atmospheric sources. Increase usage of heavy metals in some industrial, agricultural, domestic and technological applications has also increased human exposure to heavy metals (He, et al., 2005; Tchounwou, et al., 2014).

In general, high concentration of heavy metals in the environment result in health complications that disturb nervous, blood forming, cardiovascular, renal and reproductive systems. The medical implications of exposure to heavy metals pollution are reduced intelligence; loss of attention and abnormal behaviour; and cardiovascular disease in adults (Jarup, 2003; Timothy and Williams, 2019). Human exposure to heavy metals may occurs through three primary routes namely, inhalation, ingestion and skin absorption (Davis and Mundalamo, 2010).

Contamination by heavy metals may change the chemical composition of plants and thus affect the quality and effectiveness of medicinal plant species to produce organic products. Toxicity of heavy metal can damage the root system, disturb plants growth, enzymatic activity, stoma functions, photosynthesis activity and accumulation of other nutrient elements (Abrahams, 2002; Timothy and Williams, 2019). High concentration of heavy metals in soil can also diminish quality of soil, cause crop yield decrease and affect the quality of agricultural products and consequently affect people, animals and ecosystem health negatively (Nagajyoti, et al., 2010; Hu, et al., 2013; Timothy and Williams, 2019).

Heavy metals proliferation in body fatty tissues may affect human central nervous system as they are non- degradable and toxic in high concentration (Denier, et al., 2009; Ghanavati, et al., 2019; Kianpor, et al., 2019). It is important to monitor the concentration of heavy metals in the environment as they are toxic, persistent and non-degradable (Timothy and Williams, 2019). In order to assess the pollution level of heavy metals in the natural environmental samples such as soil, dust, water, etc. and determine their sources, pollution indices or methods are very useful (Othman, et al., 2019; Ghanavati, 2019; Kianpor, et al., 2019). This chapter provides an analysis of the heavy metal pollution of major public concern, their ways to environment, health implications and potentially used methods for pollution assessment and risk assessment.

Method

The information was gathered from the online peer reviewed journals, student papers, and books mainly from PubMed, MDPI, WHO, Research gate, Science direct, or Springer. This were peer reviewed journals, student papers and books published from 1969 to 2021. Only papers related to this chapter and published in English were scrutinized and utilised. Consideration of literature was based mainly on published articles with focus on heavy metal pollution, health implications, and potential used methods for pollution assessment around the world.

Results and Discussion

After thorough perusal of published manuscripts, student papers and books, only studies focusing on heavy metal pollution and public health were selected. The high percentage of manuscripts were concentrated on pollution and health risk assessment of heavy metals in urban environment. Some of the reviewed relevant concepts are discussed underneath here.

Heavy Metals of Major Public Health Concern

Any metallic element that is toxic is regarded as a heavy metal (Lenntech, 2004; Duruibe, et al., 2007). Metals and metalloids with atomic density greater than 5 g cm^{-3} and atomic mass that is higher than 40 are considered to be heavy metals (Duruibe, et al., 2007). Furthermore, heavy metals cannot be degraded or destroyed (Timothy and Williams, 2019). Out of all ninety naturally occurring elements fifty three are considered to be heavy metals. Among them, iron, molybdenum and manganese are vital as micronutrients. Zinc, nickel, copper, cobalt and chromium are considered toxic elements which are also vital as trace elements. Silver, arsenic, mercury, cadmium, antimony and lead have no known function as nutrients and are toxic to plants and microorganism. Elements that are persistent in all parts of the environment and of highest concern among the public are tin, thallium, chromium, nickel, mercury, manganese, cadmium, cobalt, copper and lead (Zhang, et al., 2012; Timothy and Williams, 2019). Heavy metals are found in different forms such as phosphates, oxides, silicates, hydroxides, sulphides, sulphates and organic compounds. When they are not metabolised by the body, they accumulate in the soft tissues and as a results they turn out to be toxic to human or animal health (Masindi and Muedi, 2018).

Properties of Heavy Metals

According to Rajeswari and Sailaja, (2014) and Briffa et al., (2020) heavy metals:

- Occur near the bottom of the periodic table.
- Have high densities over 5 g cm^{-3}.
- Are toxic in nature.
- Are persistent.
- Are non-degradable.

Emission of Heavy Metal Contamination on the Environment

From the time of earth's formation, heavy metals originate naturally on the surface of the earth (Briffa, et al, 2020). Furthermore, they become concentrated as a result of human caused activities (Rajeswari and Sailaja, 2014) and finally get deposited in air, soil and water (Masindi and Muedi, 2018).

Natural Emission

Natural sources of heavy metals into the environment are volcanic eruptions, rock weathering, forest fires, biogenic sources, sea-salt sprays and wind-borne soil particles (Herawati, et al., 2000; Masindi and Muedi, 2018). Volcanic activity (such as geothermal activity or magma degassing), continental weathering and forest fires contribute more to the release of heavy metals into the environment (air, soil and water), (Naggar, et al., 2018). Heavy metals such as copper; mercury; lead; arsenic; chromium; cadmium; zinc and nickel are mostly released naturally. Although heavy metals released naturally are found in traces, they have the potential to cause medical problems to human beings and mammals (Herawati, et al., 2000; Masindi and Muedi, 2018).

Anthropogenic Emission

Human activities have significantly increased the global emissions of trace metals in the surface environment (Nriagu, 1979). The release of pollutants to different environmental sections is contributed by activities such as agriculture, mining, industries, wastewater, metallurgical processes and runoffs. Heavy metals are emitted from industrial areas as a result of wind-blown dusts. Automobile exhaust, smelting, the use of insecticides and burning of fossil fuels also contribute significantly to the release of heavy metal such as lead; arsenic; copper; zinc; nickel; vanadium; mercury; selenium and tin in the environment. The need to meet the demands of large population through every day manufacturing of goods, have made human activities to be the most contributor of heavy metal pollution on the environment (He, et al., 2005; Masindi and Muedi, 2018). The rate of most heavy metals emissions into the environment as a result of anthropogenic activities surpass or equate natural emission rate (Naggar, et al., 2018).

Environmental Pollution from Heavy Metals

Environmental pollution by heavy metals has become a global concern due to the increase in usage in different activities to supply the needs of the increasing population (Masindi and Muedi, 2018). When heavy metal occurs in high concentration that leads to a harmful effect to both human and environment it is regarded as a pollutant. Heavy metals such as arsenic; cadmium; chromium; copper; lead; mercury; selenium; nickel; silver and zinc are considered pollutant and harmful whereas metals such as cesium; uranium; strontium; cobalt; manganese; molybdenum and aluminium are considered less common metallic pollutants (Mcintyre, 2003; Timothy and Williams, 2019).

Air Pollution by Heavy Metal

As a result of industrialisation and urbanisation, air pollution has become a serious universal environmental challenge (Masindi and Muedi, 2018). Air polluted with heavy metals is of a societal concern and considered form of pollution which is lethal. Particulate matters into air are released naturally via volcanic eruptions, rock weathering, dust storms and soil erosion whereas anthropogenic activities such as transportation and industrial activities releases large quantities of heavy metals into the atmosphere (Bilos, 2001; Timothy and Williams, 2019). Particulate matters can cause medical problems such as infections of respiratory tracts, cardiovascular infections, skin, eyes irritation and even untimely death. They can also contribute to environmental problems such as eutrophication, acid rain, haze and corrosion of infrastructure (Soleimani, 2018; Herawati, et al., 2000; Masindi and Muedi, 2018).

Soil Pollution by Heavy Metals

Soil pollution by heavy metals is mostly contributed by the release from activities such as industrial; wastewater irrigation; use of fertilisers, manures and insecticides; mine tailings; disposal of waste comprising metals; gasoline and paints with lead; combustion of coal and petrochemicals leakage (Musilova, 2016; Masindi and Muedi, 2018). Globally, soil pollution from heavy metal are of ecological concern due to their nature of being non-degradable and persistent in soil. As a result of their accumulation in soil, they become threat to soil microorganisms and biota (Okunola, et al., 2007; Oke and Vermeulen, 2016; Timothy and Williams, 2019).

The estimation of heavy metals concentration in soil is in the range of approximately less than one to 100,000 mg/kg. Soil functioning systems may be affected by the long term problems on the biogeochemical cycle as a results of soil pollution by heavy metals (Joshua, et al., 2015, Timothy and Williams, 2019). The presence of heavy metals in soil is a serious problem due to its habitation in food chains as it results in destruction of the entire ecosystem (Musilova, 2016; Masindi and Muedi, 2018).

Water Pollution by Heavy Metal

According to Timothy and Williams, (2019) water pollution generally refers to substances that have accumulated in water in an amount that is likely to affect people or animals. At higher amount, all metals are considered toxic and their occurrence in water result to water pollution (Timothy and Williams, 2019). Too much concentration of heavy metals can have detrimental effects to the organisms. The accumulation of plutonium, mercury and lead over time in animal's body can cause serious illness as they are toxic and have no known health benefits on organisms (Lane, et al., 2011; Timothy and Williams, 2019). One of the universal leading root cause of diseases is water pollution. It has been reported to be responsible for the deceases of more than 14,000 persons on a daily basis (Timothy and Williams, 2019).

Industrialisation and urbanisation has been considered as two factors that has intensified level of water contamination by heavy metals. Runoff from urban, municipalities, and industrial areas cause accumulation of heavy metals in soil and water bodies sediments (Musilova, 2016; Masindi and Muedi, 2018). Even in their tiny traces, heavy metals in water can still be very toxic and results in serious health complications to humans and other ecosystems. Food chains and food webs signify the connections between organisms. Water pollution by heavy metals really disturbs all organisms. For example, humans as an organisms that feeds at the highest degree are more prone to health complications as a result of increase of heavy metals in the food chain (Lee, et al., 2002; Masindi and Muedi, 2018).

Potential Exposure Pathways

When heavy metals are not metabolised by the body they accumulate in soft tissues and as a result they turn out to be toxic (Masindi and Muedi, 2018). Health risks from heavy metals exposure are reliant on both dose and rate of exposure (Davies and Mundalamo, 2010). Humans can become exposed to heavy metals in soil, dust or water through several routes such as inhalation, ingestion or skin contact. Exposure pathways to heavy metals permit the targeting of various body tissues or organs such as lungs and the bloodstream system (WHO, 2010; Davis and Mundalamo, 2010; Sepadi, 2019).

- Inhalation: take place as a results of inhalation of particulate matter from mining, agricultural activities, residential or industrial sites.
- Ingestion: comes about through eating of food with high concentrations of heavy metals. Through intentional or unintentional ingestion of earth materials with high level of heavy metals (Davies and Mundalamo, 2010).
- Dermal or skin contact: emerges as a result of contact with minerals, soil or water contaminated by heavy metals (Davies and Mundalamo, 2010).

Toxicity and Health Issues from Heavy Metal Pollution of Major Public Concern

Exposure to some heavy metals has been associated to a huge variety of adverse health effects, including cancer. Moreover, although some elements are essential for humans, they can be dangerous at relatively high exposure levels (Nadal, 2005). Considering their potential toxicological importance, heavy metal pollution of major public concern include: Cadmium; Mercury; Lead; Chromium; Zinc and Arsenic.

Cadmium

Cadmium at very low concentration is toxic. Long term human exposure of cadmium may cause renal dysfunction. Inhalation of dusts and fumes containing cadmium can cause lung illness and cadmium pneumonitis. As a results of lung infections from cadmium exposure, excessive accumulation of watery fluids may develop and cause death of lung tissue, chest pains, cough with foams and mucus with blood. Cadmium may also increase blood pressure, cause bone weaknesses, unprompted fractures, osteomalacia, osteoporosis and myocardic dysfunctions. Effects of cadmium exposure depends on the level of exposure and may result in symptoms such as nausea, vomiting, abdominal cramps, dyspnea and muscular weakness. Furthermore, the end results of severe exposure to cadmium are pulmonary odema effect and mortality. Respiratory and renal problems may occur after sub-chronic inhalation exposure to cadmium and its compounds (Young, 2005; Duruibe et al., 2007).

Mercury

Medical problems linked with mercury exposure are both acute and chronic. Acute exposure my cause excessive salivation, ataxia and abnormal reflexes in children; pharyngitis; dysphagia; abdominal pain; nausea and vomiting; bloody diarrhoea and shock; swelling of the salivary glands; stomatitis; loosening of the teeth; nephritis and hepatitis. Chronic exposure of mercury may results in increased excitability, irritability, psychiatric disturbances, and vibrations. Furthermore, health implications of mercury exposure mostly cause damage to neurological and renal system. However, problems that affect the lungs; kidneys; cardiovascular and immune systems; vision and hearing; that cause paralysis, insomnia and emotional instability may develop as a result of mercury exposure. Unplanned abortion, pregnancy problems in women, acrodynia or pink disease may also happen as a results of mercury exposure (WHO, 2014).

Lead

It is one of the most remarkable poisonous metal among the heavy metals. Exposure to the inorganic forms of lead mostly occurs through food and water ingestion and inhalation of particulate matters. A remarkably severe consequence of lead toxicity is its teratogenic effect. Lead exposure may cause severe and prolonged impairment to nervous system; cause inhibition of the synthesis of haemoglobin; kidney dysfunctions; dysfunction of joints and reproductive systems (Ogwuebgu and Muhanga, 2005). Furthermore, lead may cause impairment to the gastrointestinal tract and urinary tract causing bloody urine; neurological disorder; severe and permanent damage to the brain. In children, lead may cause deprived development of brain grey matter which causes reduced intelligent quotient (IQ), (Udedi, 2003; Duruibe et al., 2007). Psychosis may also occur as results of acute and chronic effects of lead (Lenntech, 2004; Duruibe et al., 2007).

Chromium

Chromium metal and chromium (III) compounds are regarded as non-health hazard (Rajeswari and Sailaja, 2014). Non-carcinogenic effects on the liver, kidney, gastrointestinal and immune systems may occur as a results of chronic exposure to Cr (VI), (Nadal, 2005). Inhalation of high concentration of chromium (VI) can results in irritation to nose lining and even nose ulcers. One of the notable health implications in animals as a results of ingestion of chromium (VI) compounds are stomach irritation; stomach and small intestine ulcers; damage to sperm and reproductive systems particularly in male and anaemia. However, such health implications do not occur due to exposure to chromium (III) compounds as it is much less toxic.

Chromium (VI) or chromium (III) exposure may cause complex reactions and allergies to some persons which may results in severe inflammation and skin swelling. Ingestion of chromium (VI) in drinking water has been reported to cause stomach cancer in both animals and humans. Deliberate or inadvertent ingestion of very high doses of chromium (VI) compounds by human beings may results in severe respiratory, cardiovascular, gastrointestinal, hematological, hepatic, renal and neurological effects or even death (Tchounwou, et al., 2014). The ability of chromium to cause cancer in human and terrestrial mammals is very strong however the manner in which it results in cancer is entirely not understood (Chen, et al., 2009; Tchounwou, et al., 2014). A value of 3×10^{-3} mg/kg/day has been recommended as chromium oral reference dose in the drinking water (Nadal, 2005).

Zinc

Ingestion of zinc is regarded to be quite non-toxic. Zinc is one of an essential element vital for regulating biochemical and physiological functioning of tissues (Tchounwou, et al., 2014). Zinc play a fundamental part in growth regulation, cell production and maintenance of stability. Lack of zinc may cause cells to die (Brieffa, et al., 2020). High concentration intake of zinc by human may cause system dysfunctions that result in growth impairment and weakening of reproduction system (Duruibe, et al., 2007). Furthermore, health problems associated with zinc toxicities are vomiting; diarrhoea; bloody urine; icterus or yellow mucus membrane; liver failure; kidney failure and anaemia (Duruibe, et al., 2007).

Arsenic

Symptoms of arsenic poisonousness depend on the chemical form that an organism is exposed to. Arsenic acts to coagulate protein, forms complexes with coenzymes and inhibits the production of adenosine triphosphate in the course of breathing. It is regarded as cancer-causing in compounds of all its oxidation states and may results in death at high level of exposure. Arsenic toxicity may also cause a condition which is similar to Guillain-Barre syndrome. Guillain-Barre syndrome is an anti-immune disorder that transpires when the body's immune system incorrectly attacks part of the peripheral nervous system which causes nerve swelling thus leading to muscle weakness (Duruibe, et al., 2007).

Heavy Metal Pollution Assessment Methods

Continuous deposition of heavy metals into soil and sediments is highly influenced by the results of industrialisation and economic development. Channels such as composting; metal mining; refining and refinishing by products; rivers; irrigation; runoff; and atmospheric deposition, introduce heavy metals to soil and sediments. The assessment of heavy metals and degree of contamination in soil and sediments needs pre-anthropogenic information of concentration of heavy metals in order to act as the original values (Barbreiri, 2016; Timothy and Williams, 2019). Pollution indices play a crucial role in effective evaluation of soil pollution with heavy metals (Kowalska, et al., 2018). For effective assessment of soil pollution, environmental quality, decision making and spatial planning, different contamination assessment methods or pollution indices have been developed based on their different procedures (Cheng, et al., 2007; Qingjie, et al., 2008).

Pollution indices are a powerful tool for assessing, processing, analysing and conveying raw environmental information to decision makers, managers, technicians and the public (Caeiro, et al., 2005; Qingjie, et al., 2008). They have the ability to monitor soil quality and ensure future sustainability (Ogunkunle and Fatoba 2013; Kelepertzis 2014; Ripin, et al., 2014; Kowalska, et al., 2018). Possible pollution assessment methods commonly used in the literature for the assessment of heavy metal pollution include but not limited to: geo-accumulation index; contamination factor; pollution load index; enrichment factor; ecological risk factor; multi-element contamination; specific pollution index and generic diatom index.

Geo-Accumulation Index (Igeo)

The geo-accumulation index (Igeo) allows to evaluate pollution level of metal contamination (Mu¨ller 1969; Oke and Vermeulen, 2016; Kowalska, et al.,

2018; Timothy and Williams, 2019). According to Ma and Singhirunnusorn (2012), the index of geo-accumulation is commonly applied in the contamination assessment by matching the levels of heavy metal found to the crustal average or background levels originally used with bottom sediments. It is computed by the following equation (1):

$$Igeo = \log_2 [Cn/1.5\ Bn] \tag{1}$$

where Cn signifies the measured concentration of the metal of interest and Bn represents the geochemical background value or crustal average. 1.5 is the constant which is presented to reduce the effect of possible variations in the background values which may be attributed to lithological differences in the sediments or soil (Ma and Singhirunnusorn, 2012). Geo-accumulation values are useful in dividing soil into different quality classes or clusters (Mu¨ller, 1969; Ma and Singhirunnusorn, 2012; Nowrouzi and Pourhabbaz 2014; Oke and Vermeulen, 2016; Kowalska, et al., 2018).

Table 1.1. Igeo classifications (Kowalska, et al., 2018)

Igeo value	Class	Soil quality
Igeo ≤ 0	0	Uncontaminated
0< Igeo ≤ 1	1	Uncontaminated to moderately contaminated
1 < Igeo ≤ 2	2	Moderately contaminated
2< Igeo ≤ 3	3	Moderately to heavily contaminated
3< Igeo ≤ 4	4	Heavily contaminated
4< Igeo ≤ 5	5	Heavily to extremely contaminated
Igeo > 5	6	Extremely contaminated

Contamination Factor (CF)

Contamination factor is helpful in the assessment of soil pollution. It permits the evaluation of soil pollution by taking into consideration the content of heavy metal in the soil and pre-industrial reference levels or background values (Kowalska, et al., 2018). The contamination factor (CF) is calculated by the equation (2) below:

$$CF = C\ sample / C\ background \tag{2}$$

where C sample represents the determined metal concentrations and C background is the metal background concentration. The contamination factor values are categorised into four different clusters, CF < 1 indicates low contamination; 1 ≤ CF < 3 indicates moderate contamination; 3 ≤ CF ≤ 6 describes considerable contamination and CF > 6 describes very high contamination (Mmolawa, 2011; Addo, et al., 2012).

Pollution Load Index (PLI)

Pollution load index is used to assess the degree of soil contamination. It provides a simple procedure to verify how soil conditions has deteriorated due to the increase of heavy metals (Varol 2011; Taofeek and Tolulope, 2012; Kowalska, et al., 2018). The pollution load index (PLI) is computed by the following equation (3):

$$PLI = n \sqrt{} (CF1 \times CF2 \times CF3 \times \ldots\ldots\ldots CFn) \quad (3)$$

where n represents the number of metals analysed and CF connotes the contamination factor computed by the equation (2). The pollution load index offers easy but fundamental ways to assess the quality of site as described in Table 1.2 (Kowalska, et al., 2018).

Table 1.2. Pollution load index (PLI) values (Kowalska, et al., 2018)

PLI Value	Category
0 < PLI ≤ 1	Unpolluted
1 < PLI ≤ 2	Moderately to unpolluted
2 < PLI ≤ 3	Moderately polluted
3 < PLI ≤ 4	Moderately to highly polluted
4 < PLI ≤ 5	Highly polluted
5 < PLI	Very highly polluted

Enrichment Factor (EF)

Originally it was established to predict the source of elements in the atmosphere, precipitation, or seawater (Duce, et al., 1975; Qingjie, et al., 2008). It is now commonly used to predict or speculate the origin or sources of pollution in soils, lake sediments, peat, tailings, and other environmental materials (Reimann and de Caritat, 2005; Qingjie, et al., 2008). The formula to calculate EF is:

$$EF = (E/R)\ sample\ /\ (E/R)\ background \quad (4)$$

where E represents the concentration of an element and R denotes a reference element of crustal material. Therefore, E/R sample is the concentration ratio of E to R in the collected samples and E/R background is the concentration ratio of E to R in the earth's crust. According to Ghavanati, et al., (2019), EF values $0.05 \leq EF \leq 1.5$ show that the toxic metal originated completely from crustal materials or a natural source, while values greater than 1.5 would show anthropogenic source of toxic metals. Degree of pollution can be categorised in five classes (Zhang and Liu, 2002; Qingjie, et al., 2008; Sebaiwa, 2016; Ghavanati, et al., 2019).

Table 1.3. Enrichment categories of EF values (Ghavanati, et al., 2019)

EF Value	Category
EF < 2	Depletion to minimal enrichment
EF = 2-5	Moderate enrichment
EF = 5-20	Significant enrichment
EF = 20-40	Very high enrichment
EF > 40	Extremely high enrichment

Ecological Risk Factor (Er^i)

An ecological risk factor (Er^i) to quantitatively express the potential ecological risk of a given contaminant was suggested by (Håkanson, 1980; Qingjie, et al., 2008). Ecological risk factor is useful in the evaluation of the degree of ecological risk triggered by heavy metal concentrations in different environmental compartment such as soil, water, and air, (Kowalska, et al., 2018; Alsafran, 2021). It is calculated by equation (5) below:

$$Er^i = Tr^i \times C^i_f \quad (5)$$

where Tr^i refers to the toxic response factor for a particular element and C^i_f denotes the contamination factor. Ecological risk factor is described by different terminologies: $Er^i < 40$ describes low potential ecological risk; $40 \leq Er^i < 80$ describes moderate potential ecological risk; $80 \leq Er^i < 160$ describes considerable potential ecological risk; $160 \leq Er^i$ describes high potential ecological risk; while $Er^i \geq 320$ describes very high ecological risk (Qingjie, et al., 2008).

Multi-Element Contamination (MEC)

Multi-element contamination index was established by Adamu and Nganje in (2010) to provide a means to measure pollution grounded on the concentration of heavy metals in soil (Kowalska, et al., 2018). Multi-element contamination concentration or values higher than 1.0 indicates heavy metal concentration in soil as a result of an anthropogenic sources. Multi-element contamination concentration <1.0 indicates heavy metal concentration from natural origin. It is computed by equation (6) below:

$$MEC = (C_1 / T_1 + C_2 / T_2 + C_3 / T_3 + \ldots C_n / T_n) / n \quad (6)$$

where C denotes the content of heavy metal, T represents the tolerable levels and n signifies the number of heavy metals (Adamu and Nganje, 2010; Kowalska, et al., 2018).

Specific Pollution Index (SPI) and Generic Diatom Index (GDI)

Specific pollution index and generic diatom index are two diatoms indices which are commonly used to gather data on nutrients, acidification, eutrophication, organic pollution and general water quality. According to Kelly, et al., (2007), as cited by Matlala (2010), these methods are good in evaluating the concentration of contamination in water. However, they are unable to measure ecological status. SPI and GDI are described as the average mean of the water quality optima (which is the tolerance limits of diatoms to water quality variables) of the texa in the sample, weighted by the quantity of each taxon. They are both grounded on the weighted average mean of the Zelinka-Marvan equation (1961) as presented in equation (7):

$$Index = \sum^{n}_{j} = 1^{ajsjvj} / \sum^{n}_{j} = 1^{ajvj} \quad (7)$$

where aj represents the proportion or an abundance of species j in a sample, vj denotes an indicator value and sj signifies a pollution sensitivity of species j. These indices are categorised into five sensitive groups as indicated in Table 1.4 below (Matlala, 2010):

Table 1.4. Evaluation of water quality using specific pollution sensitive index and generic diatom index (Matlala, 2010)

Water Quality Class	State	IPS	GDI
I	Very good	>17	>17
II	Good	15-17	14-17
III	Moderate	12--15	11--14
IV	Poor	12--8	11--8
V	Bad	<8	<8

Health Risk Assessment Method

The health risk assessment model is used to assess the health risks associated with trace elements exposure in the environment (soil, water or air) for both children and adult. Exposure pathways to heavy metals may occur through ingestion, inhalation, and dermal contact (Qadeer, et al., 2020). Non-carcinogenic and carcinogenic risk of trace elements in various environmental compartment through these exposure pathways may be assessed.

Non-Cancer Risk Assessment

To assess the non-cancer health risks, the average daily dose (ADD) of each analysed heavy metals through ingestion, inhalation and dermal contact is calculated (Gabarron, et al., 2017). It is computed by the use of equations (8) – (10) (USEPA, 1996, Gabarron, et al., 2017; Qadeer, et al., 2020).

$$ADD_{ing} = C \times IngR \times CF \times EF \times ED / BW \times AT \quad (8)$$

$$ADD_{inh} = C \times InhR \times EF \times ED / BW \times AT \times PEF \quad (9)$$

$$ADD_{derm} = C \times SA \times CF \times SL \times ABS \times EF \times ED / BW \times AT \quad (10)$$

where ADD_{ing} signifies the average daily ingestion (mg/kg/day) exposure amount of an element, ADD_{inh} indicates the average daily inhalation (mg/kg/day) exposure amount of an element, and ADD_{derm} specifies the average daily dermal (mg/kg/day) exposure amount of metal. The values of these factors are presented in Table 1.5. Non-carcinogenic risk is then evaluated from the hazard quotient (HQ) for every trace element. It was computed by dividing the ADD calculated in equations (8), (9), and (10) by a particular reference dose (RfD) as shown in equation (11):

$$HQ = ADD / RfD \tag{11}$$

where ADD less than the RfD signifies no possibility of health effects. HQ>1 suggested possibility of health effects while HQ<1 is a sign of no possibility of health effects (Yalala, 2015). The hazard index (HI) is then calculated by adding the HQ of the three various forms of exposure pathways for a corresponding element (Zgłobicki, et al., 2021). It is computed by the equation (12):

Table 1.5. Exposure factors for dose models (Qadeer, et al., 2020)

Items	Parameter	Meaning	Unit	Value	
				Children	Adult
Basic parameter	C	Concentration of a metal	mg/kg		
	D	Daily dose	mg/kg		
	CF	Conversion factor	kg/mg	1 x 10^{-6}	1 x 10^{-6}
	ED	Exposure duration	years	6	24
	BW	Body weight	Kg	15	55.9
Exposure behavioural parameter	EF	Exposure frequency	days/year	350	350
	AT	Average time (carcinogen)	days	365 × 70	365 × 70
		Average time (non-carcinogen)	days	365 x ED	365 x ED
Digestive tract/inhalation	InhR	Inhalation rate	m^3/kg	5	20
	IngR	Ingestion rate	mg/kg	200	100
	PEF	Particle emission factor	m^3/kg	1.32 x 10^9	1.32 x 10^9
Skin contact	SL	Skin adherence factor	mg/cm^2	1	1
	SA	Skin surface area	cm^2	1800	5000
	ABS	Dermal absorption	-	0.001	0.001

$$HI = (HQ)\ ing + (HQ)\ inh + (HQ)\ derm \quad (12)$$

HI value < 1 describes very low risk, HI value between 1 and 4 shows that the risk effects was possible, and HI value > 4 describes high risk (Zgłobicki, et al., 2021).

Cancer Risk Assessment

The lifetime average daily dose (LADD) of each analysed elements was also calculated for all three potential pathways of exposure (ingestion, inhalation and dermal) using equations (13) – (15), (US EPA, 2002, 1996; Ferreira-Baptista and De Miguel, 2005; Qadeer, et al., 2020).

$$LADD_{ing} = C \times CF \times EF / AT \times (IngR_{child} \times ED_{child} / BW_{child} + IngR_{adult} \times ED_{adult} / BW_{adult}) \quad (13)$$

$$LADD_{inh} = C \times EF / AT \times PEF \times (InR_{child} \times ED_{child} / BW_{child} + InhR_{adult} \times ED_{adult} / BW_{adult}) \quad (14)$$

$$LADD_{derm} = C \times CF \times EF \times SL \times ABS / AT \times (SA_{child} \times ED_{child} / BW_{child} + SA_{adult} \times ED_{adult} / BW_{adult}) \quad (15)$$

where, $LADD_{ing}$ connotes the lifetime average daily ingestion (mg/kg/day) exposure amount of a metal, $LADD_{inh}$ implies the lifetime average daily inhalation (mg/kg/day) exposure amount of an element, and $LADD_{derm}$ indicates the lifetime average daily dermal (mg/kg/day) exposure amount of a metal. Table 1.5 summarised the aspects of exposure for the above models (USEPA, 2002; Ferreira-Baptista and De Miguel, 2005; Li, et al., 2013; Lu, et al., 2014; Gabarron, et al., 2017;Qadeer, et al., 2020). After calculating the LADD of each exposure pathway, a lifetime cancer risk (CR) is then computed by multiplying the LADD with an equivalent slope factor (SL).

$$CR = LADD \times SF \quad (16)$$

The permissible risk usually range from 10^{-6} to 10^{-4} (USEPA, 1991; Rendell and McGinty, 2007; Lu, et al., 2014; Yalala, 2015; Han, 2017).

Conclusion

Heavy metal pollution is problematic globally due to their indestructible and toxic features. In abnormal concentration, heavy metals become toxic to human, animals and plants. Heavy metals occur as a results of both natural and anthropogenic processes. Natural sources of heavy metals are volcanic eruptions, rock weathering, forest fires, biogenic sources, sea-salt sprays and wind-borne soil particles. Anthropogenic sources of heavy metals include coal and fuel combustion; waste disposal; traffic emission; industrial and energy production. Continuous monitoring and assessment of heavy metals pollution levels is vital. The use of pollution indices play a crucial role in the effective monitoring and assessment of heavy metal pollution in the environment. Pollution indices are useful in finding or establishing grounds for correct interpretation of environmental conditions. To properly interpret and comprehend the level and sources of contamination in the environment, the use of proper indices is fundamental.

References

Abrahams, Peter, W. 2002. "Soils: Their implications to human health." *Sci. Total Environ.* Vol. 291, 1–32. doi.10.1016/s0048-9697 (01)01102-0.

Adamu, Christopher. L., Nganje, Therese. N., 2010. "*Heavy metal contamination of surface soil in relationship to land use patterns: A case study of Benue State, Nigeria. Materials Sciences and Applications.*" Vol. 1(3), 127–134.doi.10.4236 /msa.2010.13021.

Addo, Moses, A., Darko, Emmanuel, O., Gordon, Christopher, Nyarko, Benjamin. J. B., Gbadago, Joseph, K., 2012. "Heavy Metal Concentrations in Road Deposited Dust at Ketu-South District, Ghana." *International Journal of Science and Technology*. Vol. 2(1), 28-29, ISSN 2224-3577.

Adeola, Alex, A., Kelechi, Longinus, N., Modupe, Olatunde, A., 2015. "Assessment of Heavy Metals Pollution in Soils and Vegetation around Selected Industries in Lagos State, Nigeria." *J. of Geoscience and Environment protection*, Vol. 3(7), 11-19. http://dx.doi.org/10. 4236/gep.2015.37002.

Alsafran, Mohammed, Usman, Kamal, Al Jabri, Hareb, Rizwan, Muhammad, 2021. "Ecological and Health Risks Assessment of Potentially Toxic Metals and Metalloids Contaminants: A Case Study of Agricultural Soils in Qatar." *Toxics*, Vol. 9, 35. https://doi.org/ 10.3390/toxics9020035.

Barbreiri, Maurizio, 2016. "The Importance of Enrichment Factor (EF) and Geoaccumulation Index (Igeo) to Evaluate the Soil Contamination." *J Geol Geophys* Vol. 5, 237. doi: 10. 4172/2381-1000237.

Briffa, Jessica, Sinagra, Emmanuel, Blundell, Renald, 2020. "Heavy metal pollution in the environment and their toxicological effects on humans." *Heliyon*, https://doi.org/10.1016/j.heliyon.2020.e04691.

Bilos, Claudio, Colombo, Juarn, C., Skorupka, Carlos. N., Rodriguez-Presa, Maria, J., 2001. "Source, distribution and variability of airborne trace metals in La Plate City area, Argentina." *Environmental Pollution*, Vol. 111, 149-159.doi.org//10/S0269-7491(99)00328-0.

Caeiro, Sandra, Costa, Maria, H. B, Ramos, Tomas, B., Fernandes, F., Silveira, N., Coimbra, A. N., Painho, M., Medeiros, Guilherme, F. 2005. "Assessing Heavy Metal Contamination in Sado Estuary Sediment: An Index Analysis Approach." *Ecological Indicators*, Vol. 5 (2), 51–169.doi.10.1016/j.ecolind.2005.02.001.

Cheng, Jie. L., Shi, Zhou, Zhu, You, W., 2007. "Assessment and Mapping of Environmental Quality in Agricultural Soils of Zhejiang Province, China." *Journal of Environmental Sciences*, Vol. 19(1), 50–54.doi.10.1016/s1001-0742(07)60008-4.

Chen, Tani, L., Wise, Sandra, S., Kraus, Scott, Shaffiey, Fariba, Levine, Kaitlynn, M., Thompson, Douglas, W., Romano, Tracy, O'Hara, Todd, Wise, John, P., 2009. "Particulate hexavalent chromium is cytotoxic and genotoxic to the North Atlantic right whale (Eubalaena glacialis) lung and skin fibroblasts." *Environ Mol Mutagenesis*, Vol. 50(5), 387–393.doi.10.1002/em.20471.

Davies, Theo. C., and Mundalamo, Humbulani, R., 2010. "Environmental health impacts of dispersed mineralisation in South Africa." *Journal of African Earth Science*, Vol. 58, 652–666. doi:10.1016/j.jafrearsci. 2010.08.00.

Denier, Xavier, Hill, Elisabeth, M., Rotchell, Jeanette, Minier Christophe, 2009. "Estrogenic activity of cadmium, copper and zinc in the yeast estrogen screen." *Toxicol In Vitro*. Vol. 23(4), 569–73.doi: 10.1016/j.tiv.2009.01.006.

Duce, Robert, A., Hoffmann, Gerald, L., Zoller, William, H., 1975. "Atmospheric Trace Metals at Remote Northern and Southern Hemisphere Sites: Pollution or Natural Science." *Science*, Vol. 187(4171), 59–61.doi.10.1126/science.187.4171.59.

Duruibe, Joseph, O., Ogwuegbu, Martin, O. C., Egwurugwu, Jude, N., 2007. "Heavy metal pollution and human biotoxic effects." *International Journal of Physical Sciences*. Vol. 2 (5), 112-118.

Ferreira Baptista, L. and De Miguel, Eduardo, 2005. "Geochemistry and risk assessment of street dust in Luanda, Angola: A tropical urban environment." *Atmos. Environ.* Vol. 39(25), 4501–4512.doi.org/10 .1016/j.atmosenv.2005.03.026.

Gabarron, Maria, S., Faz, Angel, C., Acosta, Josea, A.A., 2017. "Effect of different industrial activities on heavy metal concentration and chemical in topsoil and road dust." *Environ. Earth Sci.* Vol. 76, 129. Doi.org/10.1007/s12665-017-6449-4.

Ghanavati, Navid, Nazarpour, Ahad, Watts, Michael, J., 2019. "Status, source, ecological and health risk assessment of toxic metals and polycyclic aromatic hydrocarbons (PAHs) in street dust of Abadan, Iran." *Catena*, Vol. 177, 246-259.doi.org/10.1016/j.catena.2019.02. 022.

Ghanavati, Navid, Nazarpour, Ahad, De Vivo, Benedetto, 2019. "Ecological and human health risk assessment of toxic metals in street dusts and surface soils in Ahvaz, Iran." *Environ Geochem Health*. Vol. 41(2), 875–91.doi: 10.1007/s10653-018-0184-y.

Hakanson, Lars 1980. "An ecological risk index for aquatic pollution control. A sedimentological approach." *Water Res*. 1980, Vol. 14(18), 975–1001.doi.org/10.1016/0043-1354(80)90143-8.

Han, Xiufeng, Qinggeletu, Xienwei L., Wu, Yongfu, 2017. "Health Risks and Contamination Levels of Heavy Metals in Dusts from Parks and Squares of an Industrial City in Semi-Arid Area of China." *Int. J. Environ. Res. Public Health*, Vol. 14, 886. doi:10.3390/ijerph 14080886.

He, Zhenli, L., Yang, Xiaoe, E., Stoffella, Peter, J., 2005. "Trace elements in agroecosystems and impacts on the environment." *Journal of Trace Elements in Medicine and Biology*. Vol. 19(2–3):125-140.doi.10.1016/j.jtemb.2005.02.010.

Herawati, Netti, Suzuki, Shosuke, Hayashi, Kunihiko, Rivai, Ida, F., Koyoma, Horoshi, 2000. "Cadmium, copper and zinc levels in rice and soil of Japan, Indonesia and China by soil type." *Bulletin of Environmental Contamination and Toxicology*, Vol. 64, 33-39.doi.org/10.1007/s00128991000.

Hu, Yuanan, Liu, Xueping, Bai, Jinmei, Shih, Kaimin, Zeng, Eddy, E., Cheng, Hefa, 2013. "Assessing heavy metal pollution in the surface soils of a region that had undergone three decades of intense industrialization and urbanization." *Environ Sci Pollut Res*, Vol. 20, 6150–615.doi. 10.1007/s11356-013-1668-z.

Jarup, Lars, 2003. "Hazards of heavy metal contamination." *Br. Med. Bull*., Vol. 68, 167–182.doi.org/10.1093/bmb/Idg032.

Joshua, Oluwole, O., Liziwe, Muguvhisa, L., Nomsa, Busa, B., 2015. "Original Research Trace Metals in Soil and Plants around a Cement Factory in Pretoria, South Africa." *Pol. J. Environ. Stud*. Vol. 24 (5), 2087-2093. https://doi.org/10.15244/pjoes/43497.

Kelepertzis, Efstratios, 2014. "Accumulation of heavy metals in agricultural soils of Mediterranean: Insights from Argolida basin, Peloponnese, Greece." *Geoderma*, Vol. 221–222, 82–90.9p, ref: 1p.1/4. ISSN 0016-7061.

Kelly, Martyn, G., Juggins, Stephen, Guthrie, Robin, Pritchard, Sarah, Jamieson, Jane, Rippey, Brian, Hirst, Heike, Yallop, Marian, 2007. "Assessment of ecological status in U.K Rivers using diatoms." *Freshwater Biology*, Vol. 53 (2), 403-422.doi.10.111/j.1365-2427.2007.01903.x.

Kianpor, Mandana, Payandeh, Khoshnaz, Ghanavati, Navid, 2019. "Environmental Assessment of Some Heavy Metals Pollution in Street Dust in the Industrial Areas of Ahvaz." *Jundishapur J Health Sci*., Vol. 11(3), e87212, doi: 10.5812/jjhs.87212.

Kowalska, Joanna, B., Mazurek, Ryszard, Gasiorek, Michal, Zaleski, Tomasz, 2018. "Pollution indices as useful tools for the comprehensive evaluation of the degree of soil contamination–A review." *Environ Geochem Health*, Vol. 40, 2395–2420. https://doi.org/10.1007/s10653-018-0106-z.

Lane, Todd, W., Saito, Mak, A., George, Graham, N., Pickering, Ingrid, J., Morel, Francois, M., Prince, Roger, C., 2011. "Biochemistry: A cadmium enzyme from a marine diatom." *Nature*, Vol. 435, 42.doi.10.1038/435042a.

Lee, Giehyeon, Bigham, Jerry, M., Faure Gunter, 2002. "Removal of trace metals by coprecipitation with Fe, Al and Mn from natural waters contaminated with acid mine drainage in the Ducktown Mining District, Tennessee." *Applied Geochemistry*. Vol. 17(5):569-581.doi.org/10.1016/S0883-2927(01)00125-1.

Lenntech Water Treatment and Air Purification, 2004. "*Water Treatment,*" Published by Lenntech, Rotterdamseweg, Netherlands. (www. excelwater.com/thp/filters/Water-Purification.htm). Accessed: 24/06/2021.

Li. Hulming, Qian, Xin, Hu, Wei, Wang, Yulei, Gao, Hailong, 2013. "Chemical speciation and human health risk of trace metals in urban street dusts from a metropolitan city, Nanjing, SE China." *Science of the Total Environment*, Vol. 456-457, 212-221.doi.10.1016/j. scitotenv.2013.03.094.

Lu, Xinwei, Wu, Xing, Wang, Yiwen, Chen, Hao, Gao, Panpan, Fu, Yi, 2014. "Risk assessment of toxic metals in street dust from a medium-sized industrial city of China." *Ecotoxicology and Environmental Safety*, Vol. 106,154–163.doi.10.1016/j.ecoenv. 2014.04.022.

Ma, Joshua and Singhirunnusorn, Wachitra, 2012. "Distribution and Health Risk Assessment of Heavy Metals in Surface Dusts of Maha Sarakham Municipality." *Procedia, Social and Behavioral Sciences.* Vol. 50, 280–293.doi.org/10.1016/j.sbspro.2012.08.034.

Matlala, Malebo, D., 2010. "*The use of diatoms to indicate water quality in wetlands, a South African perspective*." Magister Scientae Dissertation, University of North West, Potchefstroom campus. https://repository.nwu.ac.za/handle/10394/4410. Accessed: 27/06/ 2021.

Mcintyre, Terry, 2003. "Phytoremediation of heavy metals from soils." *Adv Biochem Eng Biotechnol*., Vol. 78, 97–123.doi.org/10.1007/3-1007/3-540-45991-x_4.

Mmolawa, Khumoetsile, B., Likuku, Alfred, S., Gaboutloeloe, Gilbert, K., 2011. "Assessment of heavy metal pollution in soils along major roadside areas in Botswana." *African J. Environmental Science and Technology*, Vol. 5(3), pp. 186-196. https://moodle. buan.ac.bw:80/ handle/123456789/181. Accessed: 16/06/2018.

Martin, Yvonne, E. and Johnson, Edward, A., 2012. "Biogeosciences survey: studying interactions of the biosphere with the lithosphere, hydrosphere and atmosphere." *Prog. Phys. Geogr.* Vol. 36, 833–852.doi.org/10.1177/0309133312457107.

Masindi, Vhahangwele and Muedi, Khathutshelo, L., 2018. "*Environmental contamination by heavy metals*." InTech.doi.10. 5772/intechopen.76082.

Musilova, Jannete, Arvay, Julius, Vollmannova, Alena, Toth, Tomsa, Tomas, Jan, 2016. "Environmental contamination by heavy metals in region with previous mining activity." *Bulletin of Environmental Contamination and Toxicology*, Vol. 97(4), 569-575.doi.10.1007/ s00128-016-1907-3.

Mu¨ller, G., 1969. "Index of geoaccumulation in sediments of the Rhine River." *GeoJournal*, Vol. 2, 108–118.

Nadal, Lomas, M. 2005. "*Human health risk assessment of exposure to environmental pollutants in the chemical / petrochemical industrial area of Tarragona (Catalonia, Spain)*." European PhD Thesis. Rovira i Virgili University. ISBN: 9788469036280. https:tdx.ac/ handle/1080/8719#page=1. Accessed: 20/06/2021.

Nagajyoti, P. C., Lee, Kap, D. and Sreekanth, Tellamekala, V. M., 2010. "Heavy metals, occurrence and toxicity for plants: a review." *Environ Chem Lett*, Vol. 8 (3): 199–216.doi.10.1007/s10311-010-0297-8.

Naggar, Yahya, Khalil, Mohamed, S., Ghorab, Mohamed, A., 2018. "Environmental Pollution by Heavy Metals in the Aquatic Ecosystems of Egypt." *Open Acc J of Toxicol*, Vol. 3(1).doi: 10.19080/OAJT.2018.03.555603.

Nriagu, Jerome, O., 1979. "Global inventory of natural and anthropogenic emissions of trace metals to the atmosphere." *Nature*, Vol. 279, 409-411.doi.10.1038/279409a0.

Nowrouzi, Mohsen, Pourkhabbaz, Alireza, 2014. "Application of geoaccumulation index and enrichment factor for assessing metal contamination in the sediments of Hara Biosphere Reserve, Iran." *Chemical Speciation & Bioavailability*, Vol. 26, 99.doi.org/10. 3184/095422914x13951584546986.

Ogunkunle, Clemet, O., Fatoba, Paulo, O., 2013. "Pollution loads and the ecological risk assessment of soil heavy metals around a mega cement factory in Southwest Nigeria." *Polish Journal of Environmental Studies*, Vol. 22(2), 487–493.

Ogwuegbu, Martin, O. C, and Muhanga, W., 2005. "Investigation of Lead Concentration in the Blood of People in the Copperbelt Province of Zambia." *J. Environ*. Vol. (1), 66 – 75.

Oke, Saheed, and Vermeulen, Danie, 2016. "Geochemical modelling and remediation of heavy metals and trace elements from artisanal mines discharge." *Soil and Sediment Contamination: An International Journal*, Vol. 26(1), 84-95. doi.org/10.1080/153203 83.2017.

Okunola, Oluwole, J., Uzairu, A., Ndukwe, George, I. 2007. "Level trace metal in soil and vegetation along major and minor road in metropolitan city of Kaduna, Nigeria." *African Journal of biotechnology*, Vol. 6 (14): 1703-1709. eISSN: 1684-5315.

Othman, Murnira, Latif, Mohd, T., Matsumi, Yutaka, 2019. "The exposure of children to PM2.5 and dust in indoor and outdoor school classrooms in Kuala Lumpur city centre." *Ecotoxicol. Environ. Saf.* Vol. 170, 739–749. doi.10.1016/j.ecoenv. 2018.12.042.

Qadeera, Abdul, Saqibb, Zulfiqar, A., Ajmal, Zeeshan, Xinga, Chen, Khalila, Saira, K., Usman, Mahammad, Huanga, Yanping, Safdar Bashir, Ahmad, Zulfiqar, Ahmede, Saeed, Thebo, Khalid, H., Liua, Min, 2020. "Concentrations, pollution indices and health risk assessment of heavy metals in road dust from two urbanized cities of Pakistan: Comparing two sampling methods for heavy metals concentration." *Sustainable Cities and Society*, Vol. 52, 101953.doi: 10.1016/j.scs.2019.101959.

Qingjie, Gong, Jun, Deng, Yunchuan, Xiang, Qingfei, Wang, Liqiang, Yang, 2008. "Calculating Pollution Indices by Heavy Metals in Ecological Geochemistry Assessment and a Case Study in Parks of Beijing." *Journal of China University of Geosciences*, Vol. 19 (3), 230–241, ISSN 1002-0705.doi. 10.1016/S1002-0705(08)60042-4.

Rajeswari, Raja, T., and Sailaja, Namburu, 2014. "Impact of heavy metals on environmental pollution. National Seminar on Impact of Toxic Metals, Minerals and Solvents leading to Environmental Pollution." *Journal of Chemical and Pharmaceutical Sciences*, Vol. 3,175-181. ISSN: 0974-2115.

Reimann, Clement and de Caritat, Patrice, 2005. "Distinguishing between Natural and Anthropogenic Sources for Elements in the Environment: Regional Geochemical Surveys versus Enrichment Factors." *The Science of the Total Environment*, Vol. 337: 91–107.doi.org/10.1016/j.scitotrnv.2004.06.011.

Rendell, Edward. G. and McGinty, Kathleen. A, 2007. "*Collegeville Area Air Monitoring Report, Commonwealth of Pennsylvania Department of Environmental Protection, Collegeville Area Air Toxics Study*." www.dep.state.pa.us. Accessed: 18/06/2021

Ripin, Siti, N. M., Hasan, Sharizal, Kamal, Mohd, L., Hashim, NorShahrizan, M., 2014. "Analysis and pollution assessment of heavy metal in soil, Perlis. *The Malaysian Journal of Analytical Sciences*." Vol. 18(1)155–161. ISSN 1394-2506.

Sands, Philippe, 2003. "*Principles of International Environmental Law*." 2nd ed. London: Cambridge, 2003.doi.org/10.1017/CBO97805118 13511.

Sebaiwa, Marks, M., 2016. "*Characterisation of dust fallout around the city of Tshwane (cot), Gauteng, South Africa*." Master thesis, University of South Africa. https://uir.unisa.ac.za/handle/10500/ 20985. Accessed: 16/06/2021.

Sepadi, Maasago, M., (2019). "*Workers' risk of exposure to inhalable and respirable dust fractions at Platinum mine waste rock crusher plants: A case study of Fetakgomo-Greater Tubatse Municipality*." Masters mini-dissertation, Faculty of Health Sciences, University of Johannesburg. Uj: 34694. http://hdl.handle.net/10210/412359. Accessed: 24/06/2021.

Soleimani, Mohsen, Amini, Nasibeh, Sadeghian, Babak, Wang, Dongsheng, Fang, Luping, 2018. "Heavy metals and their source identification in particulate matter (PM2.5) in Isfahan City, Iran." *Journal of Environmental Sciences*. Vol. 72, 166-175.doi.org/10. 1016/j.jes.2018.01.002.

Taofeek, Yakeen, A. and Tolulope, Toluwande, O., 2012. "Evaluation of some Heavy Metals in Soils along a Major Road in Ogbomoso, South West Nigeria." *J. of Environment and Earth*, Vol 2(8),71-79. ISSN: 2225-0948.

Tchounwou, Paul, B., Yedjou, Clement, G., Patlolla, Antika, K., Sutton, Dwayne, J., 2014. "Heavy Metals Toxicity and the Environment." *NIH Public*, Vol. 101, 133–164. doi:10.1007/978-3-7643-8340-4_6.

Timothy, Nachana and Williams, Ezekiel, T. 2019. "Environmental Pollution by Heavy Metal: An Overview." *International Journal of Environmental Chemistry*, Vol. 3(2), 72-82. doi: 10.11648/j.ijec. 20190302.14.

Udedi, Stanley, C. 2003. "From Guinea Worm Scourge to Metal Toxicity in Ebonyi State, Nigeria." *Nigeria Magazine*. Vol. 2(2): 13–15.

USEPA, (United States Environmental Protection Agency), 1996. "*Soil screening guidance: Technical background document, Office of Soild Waste and Emergency Response. Washington DC*." https:www. epa.gov/superfund/superfund-soil-screen ing-guidance#:~:text=The %20Soil%20Screening%2. Accessed: 23/06/2021.

USEPA, (United States Environmental Protection Agency), 1991c. "*Role of the baseline risk assessment in superfund remedy selection decisions*." April 22- memorandum. Office of solid waste and emergency response. OSWER Directive 9355.0-30. https://www.epa.gov/risk/role-baseline-risk-assessment-superfund-remedy-selection-descisions. Accessed: 22/06/2021.

USEPA, (United States Environmental Protection Agency), 2002. "*Supplemental guidance for developing soil screening levels for superfund sites, Office of Emergency and Remedial Response. Washington DC*." https:epa.gov/superfund/superfund-soil-screening-guidance. Accessed: 25/06/2021.

Varol, Memet, 2011. "Assessment of heavy metal contamination in sediments of the Tigris River (Turkey) using pollution indices and multivariate statistical techniques." *Journal of Hazardous Materials*, Vol. 195, 355–364. doi.org/10.1016/j.jhaz mat.2011.08.051.

WHO (World Health Organization), 2010. "*WHO human health risk assessment toolkit: chemical hazards*." IPCS. http://www.who. int/ipcs/methods/harmonization/areas/ ra_toolkit/en. (Accessed 17 May 2021).

WHO (World health organisation), 2014. "*Chemicals of public health concern and their management in the African region*." https://www. afro.who.int/publications/chemi cals-public-health-concern-african-region-and-their-management-regional. Accessed: 24/06/2021.

Yalala, Bongani, N., 2015. "*Characterization, bioavailability and health risk assessment of mercury in dust impacted by gold mining*." PhD, Faculty of Science, University of the Witwatersrand, Johannesburg. https://wiredspace.wits.ac.za/handle/10539/20168. Accessed: 19/06/2021.

Young, Robert, A. 2005. "*Toxicity Profiles: Toxicity Summary for Cadmium, Risk Assessment Information System, RAIS*," University of Tennessee https://rais.ornl.gov/tox/profiles/cadmium.shtml). Accessed: 24/06/2021.

Zhang, Jing and Liu, Changling, 2002. "Riverine composition and estuarine geochemistry of particulate metals in China—weathering features, anthropogenic impact and chemical fluxes." *Estuarine, coastal and shelf science,* Vol. 54(6), 1051-1070.doi.org/10. 1006/ecss.2001.0879.

Zhang, Fan, Yan, Xuedong, Zeng, Chen, Zhang, Man, Shrestha, Suraj, Devkota, Lochan, P., Yao, Tandong, 2012. "Influence of traffic activity on heavy metal Concentrations of roadside farmland soil in mountainous areas." *Int. J. Environ. Res. Public Health.* 2012, 9: 1715–1731. doi: 10. 3390/ijerph9051715.

Zgłobicki, Wojciech, Telecka, Malgorzata, 2021. Heavy Metals in Urban Street Dust: Health Risk Assessment (Lublin City, E Poland). *Appl. Sci.,* Vol. 11, 4092. https://doi.org/10.3390/app11094092.

Chapter 4

Changes in Chromium Bioavailability on Resuspension of Contaminated Sediments from a Tropical Estuary

Christiane do Nascimento Monte[1,2,*], Ana Paula de Castro Rodrigues[1,3], Alexandre Rafael de Freitas[1], Bernardo Ferreira Braz[4], Aline Soares Freire[4], Renato Campello Cordeiro[1], Ricardo Erthal Santelli[4] and Wilson Machado[1]

[1]Geochemistry Department,
Universidade Federal Fluminense, Niterói, RJ, Brazil
[2]Geology Departament,
Universidade Federal do Oeste do Pará, Santarém, PA, Brazil,
[3]Marine Biology Departament,
Universidade Federal do Rio de Janeiro, Ilha do Fundão, RJ, Brazil
[4]Analytical Chemistry Department,
Universidade Federal do Rio de Janeiro, Ilha do Fundão, RJ, Brazil

Abstract

The resuspension of contaminated sediments in the water column has been recognized as an important process of metal pollutants remobilization in historically contaminated estuaries, which can change the concentration and bioavailability of these elements. The aim of this study was to assess possible changes on the geochemical behavior of

* Corresponding Author's Email: christiane.monte@yahoo.com.br.

In: Trace Metals
Editor: Oscar M. Thygesen
ISBN: 978-1-68507-797-6

chromium (Cr) caused by sediment resuspension in the area of a hypereutrophic estuary (Guanabara Bay, Brazil) that receives domestic and industrial effluents daily during the recent decades. This study evaluated bioavailability change (BC) for Cr in estuarine sediments from Iguaçu River (located within the most impacted Guanabara Bay area), in response to laboratorial sediment resuspension experiments. The responses of sediments layers from different depth intervals were compared, since dredging activities usually promote resuspension of sediments removed from variable depths. Performed evaluations on the anthropogenic interference on sediment quality also included ecological risk index (E^if) estimates. Chromium concentrations obtained using a weak acid extraction (in a 1 mol L^{-1} HCl solution) were considered as the reactive (bioavailable) Cr phase. After resuspension along different time intervals, the uppermost sediment layers showed higher Cr concentrations in comparison with the non-resuspended control sediment. Some were above the Effect Range Low (ERL) sediment quality guideline, suggesting risks of adverse biological effects. These findings indicate increased potential bioavailability of the metal after resuspension. The E^if indicated low risk for Cr in all depth interval. The combined use of risk indices can be a useful tool for a more adequate management of dredging activities, helping in the prediction of contamination risks.

Keywords: remobilization, Guanabara Bay, dredging, anthropogenic interference

Introduction

Dredging activities have been increasingly expanded, since it is critically important for harboring development in many countries. According to the International Association of Dredging Companies, dredging operations are essential to support trading in ports, respond to increasing demands for robust trade, provide new transportation facilities for responding to population growth and urbanization, support growing tourism and environment a remediation, and protect coastal areas in response to climate change and sea level rise (Wasserman et al., 2016). However, the resuspension of contaminated sediments in the water column has been shown to be an important source of dissolved contaminants in historically contaminated estuaries (Kalnejais et al., 2010; Robert, 2012; Monte et al., 2019).

The resuspension of sediments from contaminated sites increases water turbidity and contaminants may be mobilized from anoxic sediments when exposed to oxidized conditions (Collins, 1990; Wojlast, 1990; Vale et al., 1998). Therefore, sediment resuspension experiments in laboratory have been carried out in order to contribute to the assessment of ecological risks linked to resuspension events (Cotou et al., 2005; Machado et al., 2011, Monte et al., 2015; Monte et al., 2019).

However, dredging can disrupt the physical-chemical equilibrium between water and sediment, causing the release of trace metal from sediment to water column, increasing the potential bioavailability of contaminants (Egleeton and Thomas, 2004; Montero et al., 2013; Di Risio et al., 2017). At the same time, bioavailability is controlled by several geochemical processes, such as changes on the load of chemical elements, formation of chemical substances and complexes, adsorption to solid particles, reaction with sulfides (Simpson et al., 1998; Caetano et al., 2003, Maddock et al., 2007; Cantwell et al., 2008) and/or organic matter contents and Fe and Mn oxides and hydroxides formation (Guo et al., 2015, Freitas et al., 2019).

In order to test the potential changes on metals bioavailability after resuspension events, in situ and ex situ studies were developed to evaluate this geochemical mobilization. *Ex situ* evaluations include laboratory tests inducing sediment resuspension mechanically. Their use contributes for the assessment of ecological risks linked to resuspension events, simulating the effects of dr edging in different time intervals of a pre-defined mechanical agitation (Morse, 1994; Van der Berg et al., 2001; Caetano et al., 2003; Cappuyns et al., 2006; Cantwell et al., 2008; Machado et al., 2011; Acquavita et al., 2012; Fathollazadeh et al., 2015, Monte et al., 2015; Rodrigues et al., 2017, Monte et al., 2019).

In southeastern Brazil, Guanabara Bay is the one of largest bays of Brazil, with an area of 381 km^2 (Kjerve et al. 1997). Is an important coastal system, with one of the largest metropolitan regions of Brazil with 11.8 million inhabitants (IBGE 2010). The Bay receives industrial and domestic sewage, due to lack of basic sanitation in municipalities of basin drainage(SILVEIRA *et* al, 2011). Furthermore, in some rivers of hydrographic basin, are performed dredging projects, with the objective of reduce the silting, as in the case of Iguaçu River. Due to this events, among others anthropic activities and naturals process, the sediments submitted to the resuspension in water column, which can cause the remobilization of contaminants associated with them

(Morse, 1994; Machado et al., 2011;Fatollazadeh et al. 2015; Monte et al., 2015; Monte et al., 2019).

The western region of Bay is recognized as the most impacted of the system (Kjerve et al., 1997), which receives effluents of Metropolitan Region of Rio de Janeiro State, causing concern about high contamination by trace metals in sediments (Machado et al., 2002, 2010; Monte et al., 2021). This context clearly shows the risks of remobilization of these contaminants to the water column (Maddock et al., 2007; Machado et al., 2011).

A study about remobilization process of contaminants by resuspension of sediments, produced by a possible dredging, can used by governmental offices private companies which invest in the region, following the recommend parameters by Brazilian Legislation (CONAMA 454/2012). The concern about the effect of physical disorders about metals mobilization and nutrients has been widely discussed internationally (Warnken et al., 2003), however investigation about this form of contaminants release to water column and local biota has still been little performed (Egleeton & Thomas, 2004). The aim of this study was to evaluate the distribution and enrichment of Cr over three sediment profiles from Iguaçu River (Guanabara bay; RJ; Brazil) in different depths (surface, middle, and bottom), and to assess the potential bioavailability changes after resuspension of this material over 1h (where the first changes can be observed) and 24h (considering a daily variation scenario).

Material and Methods

Study Area and Sampling

The Guanabara Bay is one of the largest bays of Brazil, with an area of 381km^2. Domestic sewage contributes with 75% of its organic pollution and industrial wastes with 25%. This complex ecosystem suffers with several impacts, including contamination by toxic metals (Maddock et al., 2007; Machado et al., 2011; Monte et al., 2017; Monte et al., 2019). Previous studies of metal pollution in Guanabara Bay have pointed out that the bay is one of the most polluted aquatic ecosystems in Brazil (Rebello et al., 1986, Vandenberg and Rebello, 1986, Leal and Rebello,1993, Barrocas and Wasserman, 1993, Baptista- Neto et al. 2000, 2006, Faria and Sanchez, 2001,

Machado et al., 2002, Carreira et al., 2002, Kehrig et al., 2003, Maddock et al., 2007, Machado et al., 2011, Baptista- Neto et al., 2013.

The Iguaçu River (drainage area of 726km^2; JICA, 2003) drains the industrial complex present at the metropolitan area of Rio de Janeiro State, beside the presence of contaminants linked to agriculture (JICA, 2003). Also important to mention, the biggest oil refinery in Brazil is located in the mouth of Iguaçu River (Barbosa and Almeida, 2001). Additionally, one of the tributaries (Sarapuí River) is responsible for an important change on water quality (increase on oxygen biochemical demand, for example) of Iguaçu River due to a huge charge of untreated domestic wastes (Silveira et al., 2011)

Then, for this study, the sampling was conducted in July 2012 at the estuarine areas of Iguaçu River (Figure 1). The core was collected in a transect along the estuarine area, three sediment cores were colleted in each estuary (cores: IR at Iguaçu River;), using acrilic tubes (10cm diameter, between 30-50cm length) previously decontaminated. The depths reached with each sediment core were: IR1 - 35cm; IR2 - 30cm; IR3 - 45cm;. All cores were sectioned in 5cm intervals, the samples were stored in plastic bags and maintained frozen, except the aliquots used at the resuspension experiment, that was conducted as soon as the profiles were sectioned.

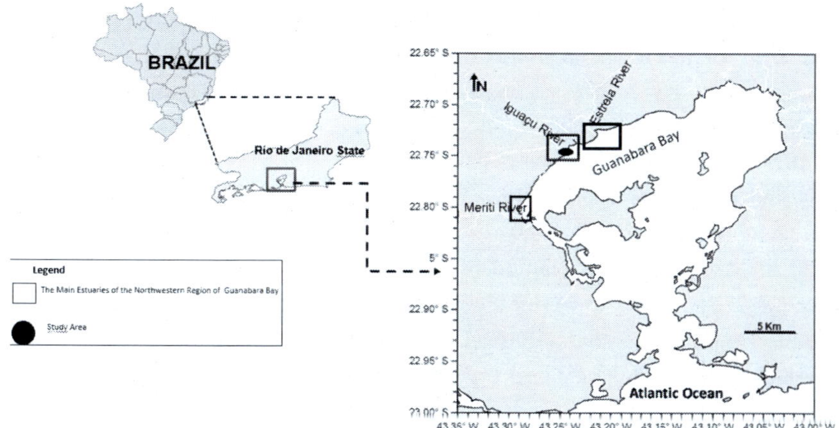

Figure 1. Study Area

According to legislation, the three sediment cores of each river became one core for each area and were used three depth (surface, middle, and bottom). For the Iguaçu River core, the surface corresponding to the 0-5 cm layer, the middle corresponding to the 15-20 cm layer, and the bottom corresponding to 30-35 cm.

Resuspension Experiment

Resuspension experiments were carried out at room temperature (25°C). The experiments compared short and long time periods of resuspension (1h vs. 24h). The intervals choice was based on previous studies, which demonstrated that in the first hour (t1) the most part of changes on bioavailibility occurs (Machado *et al.*, 2011). The second interval (t2= 24h) has been also previously adopted (Morse, 1994), and was used to evaluate the stability of the contaminant response, simulating longer resuspension events, since dregding resuspension of sediments may occur during many hours, including daily variations.

Aliquots of only three layers (surface, middle and bottom) of each sediment core were used for the experiment. These three layers were chosen with the main purpose of observe what would occur if not only the recent material (surface) was resuspended, but also older material (middle and bottom), that has different contamination degrees and phisical-chemical characteristics, which is expected to occur in cases of dredging activities. Wet sediment subsamples (7 g) were transferred to 125-mL Erlenmeyer flasks and were shaken in 100 mL of unfiltered estuarine water from Guanabara bay in contact with atmosphere. The sediment:water proportion was based on the study of Morse (1994), adapted by Machado et al. (2011). The experiments were made in duplicates. After the resuspension, the sediment was centrifuged (3,000 rpm/5min), dried (<40°C) and homogenized to posterior analysis. Eletrical condutivity (EC) and pH were measured before and after each time interval of resuspension.

Metals Determinations

The reactive phase (i.e., adsorbed and/or associated to carbonates, monosulfides and Fe and Mn oxides) was extracted by 16h agitation in 1 mol

L^{-1} HCl solution, for both non resuspended (t_0) and resuspended sediment samples. It is an usual approach for metal bioavailability evaluation (Huerta-Diaz and Morse, 1990; Morse, 1994; Machado et al., 2011; Birch and Hogg, 2011;Peña-Icart., 2014; Monte et al., 2015; Rodrigues et al., 2017). Peña Icart and co-workers (2014) compared four weak extractions (acetic acid 0.11vv, acetic acid 25% and 1M HCl with and without trypsin and pepsin) to evaluate metals bioavailability and as conclusion they indicated 1 mol L^{-1} HCl as the most efficient extraction to simulate the digestive system of marine organisms. This extraction was chosen based on Morse (1994), Machado et al. (2011) and Monte et al. (2015; 2019).

Metals concentrations (Cr, Fe and Mn) were determined by inductively coupled plasma optical emission spectrometry (ICP OES). The detection limits were: Cr: 0,01 mg Kg^{-1}; Fe: 3.00 mg Kg^{-1} and Mn: 0.02 mg Kg^{-1} and the recovery ranged from 95.49 to 107.4%.

Sediment Characterization

The sediment grain size was characterized using a particle size analyzer CILAS 1064 and calculated using the software GRADSTAT 1.0. The organic matter (OM) in sediments was determined with heating (450°C) by four hours and half.

In order to characterize the contamination and potential risks to local biota, two indexes were applied to evaluate the sediment quality of these areas. First, an enrichment factor, called by Hakanson (1980) as contamination factor (CF) were calculated in relation to mean background values obtained from ^{210}Pb-dated sediment cores sampled in Guanabara Bay (Monteiro et al., 2012). These results were classified as suggested by Hakanson (1980).

The second index was proposed by Guo and co-workers (2010) and it is derived from Hakanson (1980). It is called "potential ecological risk index for a single heavy metal" ($E^i f$ *for short*). The formula of $E^i f$ is expressed as:

$$E^i f = Cf * T^i f$$

where: Cf is the contamination factor and $T^i f$ is the response coefficient for the toxicity of the single heavy metal. Respectively, the coefficients based on its toxicity are, as proposed by Hakanson (1980): Cd=30, Cu=Pb=Ni=5, Cr=2, Zn=1. The sum of all $E^i f$ calculated separately to each metal represents the potential ecological risk caused by the overall contamination (Suresh et al., 2012), classifying areas according to the values described at Table 1.

Table 1. Grades of standard for $E^i f$ for short and PERI

$E^i f$ for short	Ecological risk Index	PERI	Grades of potential ecological risk
< 40	Low	<150	Low-grade
40-80	Moderate	150-300	Moderate
80-160	Higher	300-600	Severe
160-320	Much Higher	>600	Serious
>320	Serious		

Bioavailability Changes (BC)

The Bioavailability Change Index (BCI) developed by Monte et al. (2015) was adapted in Rodrigues et al. (2017), and include only the concentrations of metals in reactive phase (BC). Then a simple calculation of reactive trace metals losses and gains as the percentage related to t_0 (in natura samples, before resuspension) was done, as explained in the following equation: $(([Me_{AR}] - [Me_{BR}])/[Me_{BR}])*100$, where $[Me_{AR}]$ is the metal concentration on reactive phase after resuspension and $[Me_{BR}]$ is the metal concentration on reactive phase before resuspension.

Data Analysis

Statistical analysis was performed using STATISCA 7.0 software. The Principal Component Analysis (PCA) test was applied to verify possible relationships between toxic metals, grain size and organic carbon contents. This is multivariate analysis which allows for investigations with a large amount of available data (Wu et al. 2014, 2019; Li et al. 2019; Silva e Silva et al. 2020).

Results

Profiles Characterization

The sediment core of Iguacu River (IR) were mainly composed by fine particules (above 90%) (Figure 2). The percentage of organic matter (OM) was above 18% (minimum of 18.4% at the bottom; a maximum of 19.2% at the surface). The highest OM contents were in profile on the surface (Figure 3).

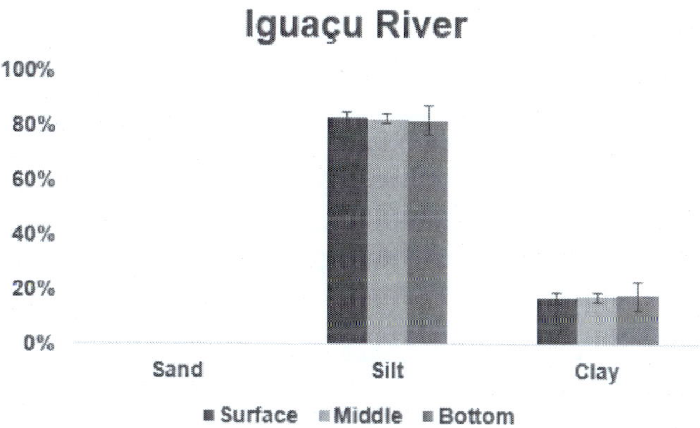

Figure 2. Grain size of cores Iguaçu River (%).

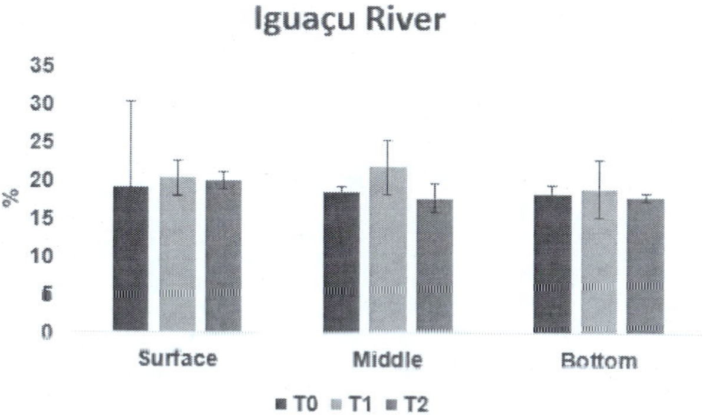

Figure 3. Organic Matter of cores of Iguaçu River (%).

Resuspension Experiment

Initially, physical-chemical changes on water from Iguaçu core experiments will be presented. The sediment core presented a change in pH considering the initial pH (Table 2) and the range observed after resuspension), showing a decreasing trend, mainly, after 24h of resuspension in all layers. Conductivity decreased after 1h of resuspension and increased after 24h in comparison to the initial water conductivity in all depths.

Table 2. Water physico-chemical in water before (t_0) and after 1 hour (t_1) and twenty-four hours (t_2) of resuspension in laboratory of core samples from Iguaçu River

Iguaçu Core						
	pH			Conductivity(ms cm^{-1})		
	T0	T1	T2	T0	T1	T2
Surface	7.8	7.6±0.1	6.0 ±0.7	76.6	73.6±0.7	77.9±0.7
Middle	7.8	7.7 ±0.1	6.4 ±0.6	76.6	73.8 ±0.7	77.8± 0.6
Bottom	7.8	7.7±0.2	6.6 ±0.1	76.6	73.9 ±0.2	78.4 ± 0.1

The concentrations of Cr, Fe, and Mn were higher in the middle layer (Figure 4), suggesting that the principal contaminant input was in an earlier period. Concerning resuspension in the surface layer, the T2 interval showed a trend of higher Cr concentrations compared to the T0 and T1 intervals (Figure 4), suggesting that the remobilization of Cr occurred after resuspension, which increases the bioavailability of the metal.

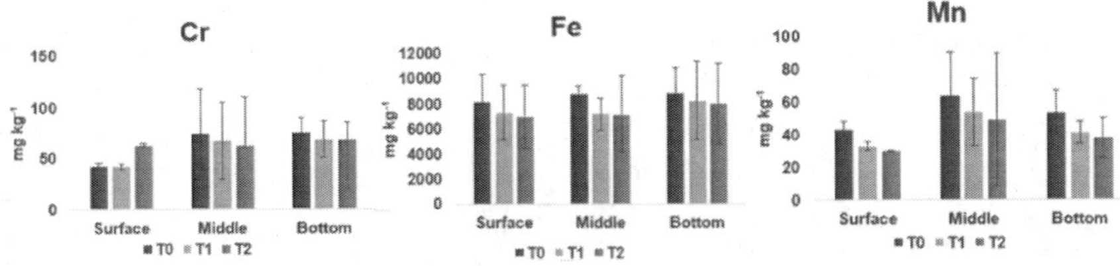

Figure 4. The concentration of Cr, Fe and Mn in profile of Iguaçu river (T0, T1 e T2) mg/Kg $^{-1}$.

The *E'f for short* showed low values for Cr, however, the highest values are in the bottom layers in all core samples (Table 3). The Bioavailability Changes (BC) (Table 4) were negatives for most samples after resuspension for sediment cores, however above 15%. The negative result in T2 suggests remobilization of these metals to water column or the influence of OM (or Fe and Mn) on metal's partitioning in a solid phase, interfering on the bioavailability of trace metals on sediments. The positive bioavailability change was in the sediment core after resuspension (T2) in the surface layer.

Table 3. E'f for short before and after resuspension in cores of Iguaçu River

	Depth cm	Time interval	Cr
I1 Core	0-5	T0	3.46
	15-20	T0	3.66
	30-35	T0	5.06
	0-5	T1	3.61
	15-20	T1	3.83
	30-35	T1	4.81
	0-5	T2	3.67
	15-20	T2	3.38
	30-35	T2	4.71
I2Core	0-5	T0	4.01
	15-20	T0	5.01
	25-30	T0	6.09
	0-5	T1	3.90
	15-20	T1	4.14
	25-30	T1	5.32
	0-5	T2	3.75
	15-20	T2	2.59
	25-30	T2	5.35
I3 Core	0-5	T0	3.43
	15-20	T0	10.80
	30-35	T0	7.61
	0-5	T1	3.43
	15-20	T1	9.66
	30-35	T1	7.77
	0-5	T2	3.34
	15-20	T2	10.21
	30-35	T2	7.70

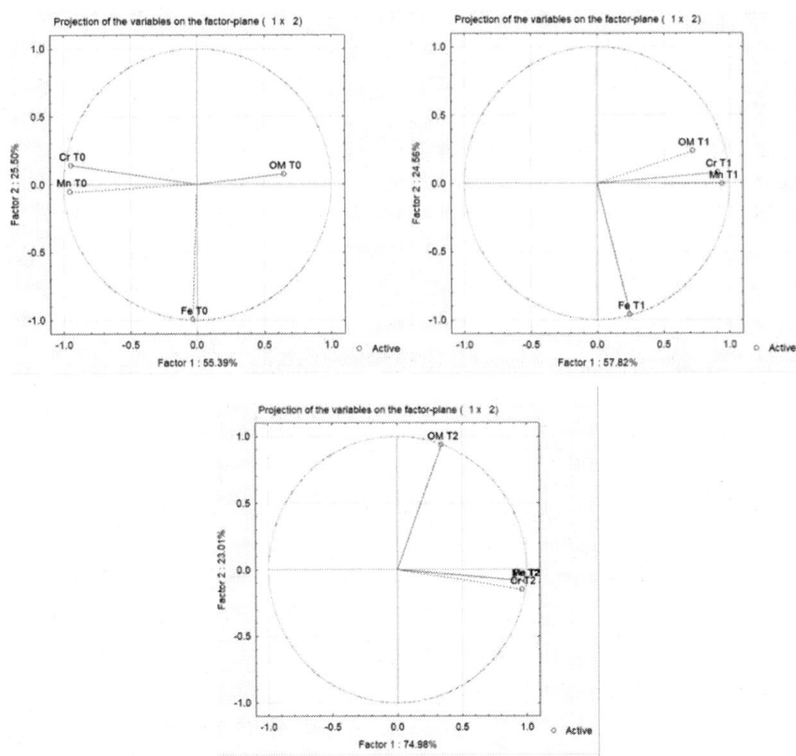

Figure 5. Plot of PCA test before and after resuspension

Table 4. Bioavailability Changes (BC) in T1 (1h) and T2 (24h) for cores of Iguaçu River (IR)

Iguaçu River			
Trace Metals	Depth	T1	T2
Cr	Surface	0.3	**48.5**
	Middle	-9.5	**-16.9**
	Bottom	-8.8	-9.5

The bold values correspond to results at least 15% different from metal concentrations in surface sediments before resuspension. This criterion was established in Monte et al. (2015).

The statistical PCA analysis (Figure 5) with two components responsible for about 75% of the results, suggested that after the T2 interval, Cr showed an affinity with Fe and Mn, being the probable regulators of bioavailability of the trace metal, which is widely discussed in the literature.

Discussion

Despite the weak extraction, the present results showed the high contamination degree of study area. Academic publications concerning Guanabara Bay pollution are current (Table 5) and includes evaluations of sediment contamination using other extraction methods, such as semi-total or total extractions.

Table 5. The average of trace metals concentrations (ppm) in surface sediments of Guanabara Bay according to the literature, considering with different extraction methods

Location	Cr mg/kg $^{-1}$
This Study Iguaçu River	42
Porto et al (2014) Caceribu River (Total extraction)	23
Porto et al. (2014) Guaxindiba River (Total extraction)	32
Pereira et al. (2007) Guanabara Bay (Semi Total extraction)	169
Baptista-Neto et al. (2006)Guanabara Bay (Total exctration)	64

The cores are composed, mainly, fine fraction (clay + silt), can contribute to adsorption of trace metals (Zhang et al. 2014). The metals sorption to sediments depends not only of ionic exchanges, but also of complexation reactions, which enable stability at bonding processes on sediments surface (Calmano et al., 1988). Thus, the organic matter has relevance in complexation of dissolved metals, which could make them less available to solution (Di Toro et al., 2005; Richards, 2012; Rodrigues, 2013; Rodrigues et al., 2017).

Some studies performed sequential extractions in coastal sediments have shown that Cr is mainly bound to the residual fraction (Chen et al., 2019; Shao et al., 2020; Ferrans et al., 2021), Cr is one of the most abundant metals in the Earth's crust, which may explain these results (Baraud et al., 2017; Ferrans et al., 2021). According to Ferrans et al. (2021), shows high affinity for organic matter, around 10% of Cr-contaminated marine sediment is bound to the organic matter-sulfide fraction (Morillo et al., 2004; Ferrans et al., 2021).

The trace metal in core presented higher concentrations of trace metals on middle and bottom layers compared to surface layers. Besides differences on sediment characteristics, such as grain size, OM, sulfides (Zhang et al., 2014), this decrease from basis to the top is linked to the changes on drainage basin throughout time.

High concentrations of Cr in the Iguaçu River region, increasing from the top to the bottom of the sediment. Industrial use, specifically in ferrous alloys and electrodeposition processes, due to its resistance to corrosion is one of the principal sources of Chromium and is therefore widely used in industries, (Bielicka et al., 2005; Aguiar et al., 2016). The regions studied present a large concentration of these industries around the rivers. However, in the region of the Iguaçu River, they present more productions related to the chromium as textile industry, of painting, oil etc.

Godoy et al. (2012), observed an increase in Cr and Cu concentrations from the inauguration of Bayer industry in 1958, in which metal effluents were released until 1982, from then on there was a decrease in their concentration in the region, due to the creation of a treatment plant. However, the levels found are still above what is considered safe for biota.

Monteiro (2011) evaluated three sediments cores of Guanabara Bay (South, East and Northwest sectors of Bay) and observed for the Northwest sector (near to Iguaçu River) a sedimentation rate of 1.9 cm year^{-1}. Applying this sedimentation rate to our sediment cores, they would register around 15 years of Iguaçu River of environmental histories. The higher concentrations of Cr, observed in intermediated layers could correspond 1990’s and 2000's decade. On the one hand, the mouth of the Iguaçu River in addition to being close to the oil refinery is close to a highway, which can contribute to Cr enrichment in the sediment, via dry deposition, due to the large vehicle traffic (Abdallah, 2014).

Baptista-Neto and co–workers. (2013) performed a study in Guanabara Bay, using one sediment core of 80 cm depth (the North sector). The core was dated with ^{210}Pb and its calculated age was 54 years, furthermore the average sedimentation rate calculated to Guanabara Bay was 0.67 cm year^{-1}. The authors described found higher fluxes of trace metals (Cr) in until 1990’s, when the Brazilian environmental regulations were implemented.

The organic matter contents in all cores were very high, demonstrating the trophic status of these areas under eutrophication process (Kjerfve et al., 1997, Machado et al., 2002; Aguiar et al.,2011). Some authors had pointed out the importance of OM on metals complexation in Guanabara Bay (Kehrig et al.,

2003, Covelli et al., 2012) and how it would influence on remobilization of trace metals (Carvalho et al., 1991). Regarding the potential bioavailability of metals, Carvalho et al. (1991) suggested that OM contents could cause the low bioavailability observed in Guanabara Bay.

However, the disposal of dredged sediment on oxidant water can cause the metals remobilization to water column and increase their bioavailability (Machado et al., 2011). Furthermore, the disposal of dredged in oligotrophic environmental can influenced on decomposition and mineralization of organic matter in this environment (Prigyel et al., 2015). In this study, generally, the cores increased the OM in some layers, suggesting primary production after resuspension.

In this study, the high percentage of OM can be influenced in few changes in remobilization of trace metals after resuspension, influencing the results of BC. The increase of percentage of OM after resuspension, can complex Cr on other geochemical phases, decreasing the reactive phase concentrations (Zhang et al., 2014). Cordeiro et al. (2015) studied regions of Guanabara Bay and the site with higher values of Cr metal was present in the mouth of the Iguaçu River, specifically in the organic fraction with percentages of 69.3%.

According to the PCA test, the organic matter had the greatest influence in the T1 interval, while in T2, the main influences were Fe and Mn. In several studies, Cr is associated with oxyhydroxides (Pazos-Capeans et al. (2010); Bastami et al., 2017; Baraud et al., 2017). According to Pazos-Capeans et al. (2010) and Baraud et al. (2017) part of the chromium probably remains in sediments as an oxide or hydroxide combined with other metals or alone or combined with organic matter as natural humic or fulvic acids. These fractions can be assimilated into oxidizable and reducible fractions.

Chen et al. (2019) performed a study of dredged sediments in a harbor area in China, and according to the results found by the authors, it was concluded that Cr was mainly bound to Fe/Mn oxyhidroxides, contrary to some studies, which report that Cr is mainly bound to the residual fraction, presenting low remobilization to other more bioavailable fractions (Morillo et al., 2004; Neyestani et al., 2016; Schintu et al., 2016; Ferrans et al., 2021). However, in dredged sediments, the behavior of Cr is different, due to the strong interference of anthropic activity, which results in higher percentages bound to Fe/Mn oxyhidroxides. Furthermore, after dredging, contact with oxygen could potentially release Cr, however, according to Ferrans et al. (2021) dredging activity isolated will not increase the bioavailability of Cr, bioavailability can increase if anthropized sediments are remobilized, as in the dredging process.

The significant results for BC were at interval T2. Although the surface layer showed lower Cr concentrations compared to the other layers, the surface layer showed an increase in BC at T2, suggesting that after resuspension for 24 hours the surface Cr becomes more potentially bioavailable, increasing the risk to biota.

Monte et al. (2015) presented a study in a mesotrophic estuary from Brazil, and the values of BCI were positives for majority samples, however with low remobilization after resuspension. In this estuary, the percentage of Total Carbon Organic (TOC) was low for majority samples due to be a mesotrophic estuary, different of Guanabara Bay, which is eutrophic estuary. Furthermore, after resuspension, the percentage of trace metals in strong bound was lower compared to before resuspension, showing change of geochemical phase. Although low percentage, these results showing the importance of OM for bioavailability of trace metals in sediments in mesotrophic estuary.

In the T2 interval, the middle layer obtained a significant negative BC result, suggesting the losses of metals in reactive phase in these sediment samples. This could be consequence of at least two major processes: (1) metals losses to water column as a result of dissolution; (2) resuspension increased the ligands possibilities and the abundance on OM could react with reactive metals and transform them in more stable substances.

Monte et al. (2019) studied surface sediments of the Iguaçu River in twelve points in the estuary after resuspension in two intervals (1 hour and 24 hours). The authors found negative values of BCI. The possible explanation for negative results for the BCI is the influence of oxidation of contaminants adsorbed to sediment, causing their mobilization to the water column, mainly at T1 (1 h).

On the other hand, can be the transference for water column, as found by Acquavita et al. (2012) and Cappuyns et al. (2006) due to sulfides oxidation, the formation Fe/Mn oxyhidroxides unstable and other process. This authors, found after resuspension, an increased in concentration of trace metals on water column, suggesting the remobilization from solid phase to dissolved phase, which may increase the risk to biota. The low remobilization of trace metals, after resuspension, in the bioavailable phase may be related to grain size. The cores had in their composition sand, which increases the bioavailability of metals in the water column (Zhang et al., 2014).

The results were close to the Effects Range Low (ERL). Can be explained by previous studies, which may showed the influence of mangrove on retention of Zn on estuary of Iguaçu River as others trace metals (Machado et al., 2002). Although the relatively low E^if for short and PERI values, however dredging activity in this estuary presents risks to biota, as was described by Monte et al. (2019).

Aguiar et al. (2016) considered the northwest portion of Guanabara Bay can be considered a hot spot of trace metal contamination, due to the low hydrodynamic and shallow depths of this location and fine granulometry. The authors considered the Bay as highly contaminated by trace metals and suggest potential risk to fishing activities.

Fernandes et al. (2020) presented the results of the cores of Guaxindiba and Caceribu river, both estuaries located in Guanabara Bay next to the Environmental Protection Area. These authors showed an elevated anthropogenic input of OM and trace metals, furthermore, the bioavailability of trace metals is linked to the soluble fraction of sediments, which represents a high risk in dredging operations due to the effects of resuspension of sediments.

Conclusion

The core sediments of Iguacu River showed clear anthropogenic influences. The possible influence of OM on results was confirmed by trends and statistical test, suggesting that OM could be an important complexing ligand in sediments of eutrophic estuaries. However, the bioavailability of dredged sediments influenced by anthropogenic activities is controlled by the formation of Fe/Mn oxyhidroxides, which showed a high affinity for Cr after resuspension.

BC showed low remobilization after resuspension, however this low remobilization, was found on weakly bound, this does not mean, which this samples has low remobilization in other fractions and water. Dredging activities represent a risk to biota, due to the resuspension of sediments, and the consequent remobilization of trace metals to the water column, due to sediment-water partition.

Other studies about resuspension in eutrophic estuaries used another geochemical fractions and ecotoxicological assays would be performed to better explain the remobilization, behavior and risks of trace metals of contaminated sediments and to calibrate the bioavailability changes calculations.

Acknowledgments

The authors would like to thank DSc. Daniel Dias and MSc. Rodrigo Carvalheira for the support during the sampling campaigns.

References

Abdallah, M.A.M. (2014). Chromium geochemistry in coastal environment of the Western Harbour, Egypt: water column, suspended matter and sediments. *Journal Coast Conservation*. 18, 1-10. https://doi.org/10.1007/s11852-013-0288-6.

Abreu, I.M; Cordeiro, R.C., Soares-Gomes, A., Abessa, D.M.S., Maranho, L.A and Santelli, R.E. (2016). Ecological risk evaluation of sediment metals in a tropical Euthrophic Bay, Guanabara Bay, Southeast Atlantic. *Marine Pollution,* 109, 435-455.

Acquavita, A., Emili, A., Covelli S, Faganeli, J., Predonzani, S., Koron, N and Carrasco, L. (2012). The effects of resuspension on the fate of Hg in contaminated sediments (Marano and Grado Lagoon, Italy): Short-term simulation experiments. *Estuarine Costal Shelf*,113:32–40.

Aguiar, V.M.C., Lima, M.N., Abuchacra, R.C., Abuchacra, P.F.F., Batista-Neto, J.A, Borges, H.V and Oliveira, V.C. (2016). Ecological risks of trace metals in Guanabara Bay, Rio de Janeiro, Brazil: An index analysis approach. *Ecotoxicology and Environmental Safety*,133 306–315.

Baptista-Neto, J.A., Smith, B.J and Mcallister, J.J. (2000). Heavy metal concentrations in surface sediments in a nearshore environment, Jurujuba Sound, Southeast Brazil. *Environmental Pollution,* 109:1–9.

Baptista-Neto, J.A., Gingele, F.X., Leipe, T. and Brehme, I. (2006). Spatial distribution of heavy metals in surficial sediments from Guanabara Bay: Rio de Janeiro, Brazil. *Environmental Geology*,49:1051–1063.

Baptista-Neto, J.A., Peixoto, T.C.S., Smith, B.J., Mcalister, J.J., Patchineelam, S.M., Patchineelam, S.R. and Fonseca, E.M. (2013). Geochronology and heavy metal flux to Guanabara Bay, Rio de Janeiro state: a preliminary study *Anais Academia Brasileira de Ciências* [*Anais Brazilian Academy of Sciences*], 85, 1317-1327.

Baraud, F., Leleyter, L., Lemoine, M. and Hamdoun, H. (2017). Cr in dredged marine sediments: anthropogenic enrichment, bioavailability and potential adverse effects. *Marine Pollution Bulletin*, 120, 303-308. https://doi.org/10.1016/j.marpolbul.2017.05.039.

Barbosa, M.C. and Almeida, M.S.S. (2001). Dredging and disposal of fine sediments in the State of Rio de Janeiro, Brazil. *Journal of Hazardous Materials*, 85,15-38.

Barrocas, P.R. and Wasserman J.C. (1993). *O mercúrio na Baía de Guanabara: um revisão histórica.* Programa de Pós graduação em Geoquímica, [*The mercury in Guanabara Bay: a historical review*. Graduate Program in Geochemistry] UFF, Niterói, RJ, p. 115-127. In Portuguese.

Bastami, K.D., Neyestani, M.R., Esmaeilzadeh, M., Haghparast, S., Alavi, C., Fathi, S., Nourbakhsh, S., Shirzadi, E.A. and Parhizgar, R. (2017). Geochemical speciation, bioavailability and source identification of selected metals in surface sediments of the Southern Caspian Sea. *Marine Pollution Bulletin*, 114, 1014–1023.

Bielicka, A., Bojanowska, I. and Wisniewskl, A. (2005). Two faces of chromium – Pollutant and bioelement. *Polish Journal Environmental Studies*, 14, 5–10.

Birch, G, F. and Hogg, T.D. (2011). Sediment quality guidelines for copper and zinc for filter-feeding estuarine oysters? *Environmental Pollution*, 159, 108-115.

Caetano, M., Madureira, M.J. and Vale, C. (2003). Metal remobilization during resuspension of anoxic contaminated sediment: short-term laboratory study. *Water, Air and Soil Pollution*, 143, 23-40.

Calmano, W., Ahlf, W. and Forstner, U. (1988). Study of metal sorption desorption processes on competing sediment. *Evironmental Geology and Water Sciences*, 11, 1, 77-84.

Cantwell, M.G., Burgess, R.M., King, J.W. (2008). Resuspension of contaminated field and formulated reference sediments Part I: Evaluation of metal release under controlled laboratory conditions. *Chemosphere,*73, 1824-1831.

Cappuyns, V., Swennen, R. and Devivier, A. (2006). Dredged river sediments: potential chemical timebombs? A case study. *Water, Air, and Soil pollution*. 171, 49-66.

Carvalho, C. E. V., Lacerda, L. D. and Gomes, M. P. (1991). Heavy metal contamination of the marine biota along the Rio de Janeiro Coast, SE Brazil. *Water, Air, and Soil Pollution*, 57/58, 645-653.

Chen, C. Ju, Y., Chen, C and Dong, C. (2019). Changes in the total content and speciation patterns of metals in the dredged sediments after ocean dumping: Taiwan continental slope. *Ocean and Coastal Management*, 181,104893 https://doi.org/10.1016/j.ocecoaman.2019.104893.

Cordeiro RC, Machado W, Santelli RE, Figueiredo AG Jr, Seoane JCS, Oliveira EP, Freire AS, Bidone ED, Monteiro FF, Silva FT and Meniconi, MFG. (2015). Geochemical fractionation of metals and semimetals in surface sediments from tropical impacted estuary (Guanabara Bay, Brazil). *Environment Earth Sciences*, 74:1363–1378.

Covelli S., Protopsalti I., Acquavita A., Sperle M., Bonardi, M. and Emili A. (2012). Spatial variation, speciation and sedimentary records of mercury in the Guanabara Bay (Rio de Janeiro, Brazil). *Continental Shelf Research*, 35, 29–42.

Di Risio., M., Pasquali, D., Lisi, I., Romano, A., Gabelini, M. and Girolam, P. (2017). An analytical model for preliminary assessment of dredging-induced sediment plume of far-field evolution for spatial non homogeneous and time varying resuspension sources. *Coastal Engineering*,127, 106-118.

Di Toro, D.M., Mcgrath, J.A., Hansen, D.J., Berry, W.J., Paquin, P.R., Mathew, R., Wu, K.B. and Santore, R.C. (2005). Predicting sediment metal toxicity using a sediment biotic ligand model: methodology and initial application. *Environmental Toxicology and Chemistry*, 24 2410–2427.

Egleeton, J and Thomas, K.V.A. (2004). Review of Factors Affecting the Release and Bioavailability of Contaminants During Sediment Disturbance Events. *Environment International*, 30,973-980.

Faria, M.M and Sanchez, B.A. (2001). Geochemistry and mineralogy of recent sediments of Guanabara Bay (NE sector) and its major rivers-Rio de Janeiro State-Brazil. *Anais Academia Brasileira de Ciências* [*Anais Brazilian Academy of Sciences*], 73: 121-133.

Fathollahzadeh, H., Kaczala, F., Bhatnagar, A. and Hogland, W. (2015). Significance of environmental dredging on metal mobility from contaminated sediments in the Oskarshamn Harbor, Sweden. *Chemosphere*, 119, 445-451.

Fernandes, M., Fonseca, E.M., Lima, L.S., Sichel, S.E., Delgado, J.F., Correa, T.R., Aguiar, V.M.C. and Baptista-Neto, J.A. (2020). Assessment of trace metal contamination and bioavailability in an Environmental Protection Area: Guaxindiba estuarine system (Guanabara Bay, Rio de Janeiro Brazil). *Regional Studies in Marine Science*, 35, 10143.

Ferrans, L., Jani, Y., Burlakovs, Klavins, M. and Hogland, W. (2021). Chemical speciation of metals from marine sediments: Assessment of potential pollution risk while dredging, a case study in southern Sweden. *Chemosphere*. 263, 128105. https://doi.org/10.1016/j.chemosphere.2020.128105

Freitas, A.R; Rodrigues, A.P.C., Monte, C.N., Soares, A.F., Santelli, R.E., Machado, W. and Sabadini-Santos, E. (2019). Increase in the bioavailability of trace metals after sediment resuspension. *SN Applied Sciences*. 1:1288 https://doi.org/10.1007/s42452-019-1276-8

Godoy, J. M., Oliveira, A. V., Almeida, A. C., Godoy, M. L. D. P., Moreira, I., Wagner, A. R. and Figueireido Junior, A. G. (2012). Guanabara Bay Sedimentation Rates based on ^{210}Pb Dating: Reviewing the Existing Data and Adding New Data. *Journal of Brazilian Chemical Society*, 23, 7, 1265-1273.

Guo, W., Liu, X., Liu, Z. and Li, G. (2010). Pollution and Potential Ecological Risk Evaluation of Heavy Metals in the Sediments around Dongjiang Harbor, Tianjin. *Procedia Environmental Sciences*, 2, 729- 736.

Guo, W., Huo, S., Xi, B., Zhang, J. and Wu, F. (2015). Heavy metal contamination in sediments from typical lakes in the five geographic regions of China: Distribution, bioavailability, and risk. *Ecological Engineering*,81,243–255.

Hakanson, L. (1980). Ecological risk index for aquatic pollution control. a sedimentological approach. *Water Research*, 14, 975 -1001.

Huerta-Diaz, M.A. and Morse J.W. (1990). A quantitative method for determination of trace metal concentration in sedimentary pyrite. *Marine Chemistry*, 29,119–144.

Islam, M.S., Ahmed, M.K., Raknnuzzman, M., Al-Mamun, M.H. and Islam, M.K. (2015).Heavy metal pollution in surface water and sediment: A preliminary assessment of an urban river in a developing country. *Ecological Indicators*,48, 282-291.

JICA. Japan International Cooperation Agency. (2003). *The study on management and improvement of the environmental conditions of Guanabara Bay of Rio de Janeiro,*

The Federative Republic of Brazil (p. 412). Rio de Janeiro, RJ: JICA and the State Secretariat of Environment and Urban Development.

Kehrig, H.A., Costa, M., Moreira, I. and Malm, O. (2003). Total and methylmercury in a Brazilian estuary, Rio de Janeiro. *Marine Pollution Bulletin*, 44, 1018-1023.

Kjerfve, B., Ribeiro., C.H.A., Dias, G.T.M., Filippo, A.M. and Quaresma, V.S. (1997). Oceanographic characteristics of an impacted coastal bay: Baía de Guanabara, Rio de Janeiro. Brazil. *Continental Shelf Research*, 17, 1609-1643.

Li P., Tian R. and Liu R. (2019). Solute geochemistry and multivariate analysis of water quality in the Guohua phosphorite mine, Guizhou Province, China. *Exposure and Health*,11(2),81–94. https://doi. org/10.1007/s12403-018-0277-y

Leal, M. and Rebello, A.H. (1993). Remobilization of anthropogenic copper deposited in sediments of a tropical estuary. *Chemical Speciation and Bioavailability*, 24(1), 31-39.

Long, E.R., Macdonald, D.D., Smith, S.L. and Calder, F.D. (1995). Incidence of adverse biological effects within ranges of chemical concentrations in marine and estuarine sediments. *Environmental Management*, 19:81–97.

Machado, W., Moscatelli, M., Rezende, L. G. and Lacerda, L.D. (2002). Mercury, Zinc, and Copper Accumulation in Mangrove Sediments Surrounding a Large Landfill in Southeast Brazil. *Environmental Pollution*, 120, 455 – 461.

Machado, W., Rodrigues, A.P.C., Bidone, E.D., Sella, S.M. and Santelli, R.E. (201). Evaluation of Cu potential bioavailability changes upon coastal sediment resuspension: an example on how to improve the assessment of sediment dredging environmental risks. *Environmental Science and Pollution Research*, 18,1033-1036.

Maddock, J.E.L., Carvalho, M.F., Santelli, R.E. and Machado, W. (2007). Contaminant Metal Behavior During Re-suspension of Sulphidic Estuarine Sediments. *Water, Air and Soil Pollution*, 181,193-200.

Monte, C.N., Rodrigues, A.P.C., Cordeiro, R.C., Freire, A.S., Santelli, R.E. and Machado, W. (2015). Changes in Cd and Zn bioavailability upon an experimental resuspension of highly contaminated coastal sediments from a tropical estuary. Sustainable *Water Resources Management*, 1, 332-335.

Monte, C.N., Rodrigues, A.P.C., Freire, A. S., Santelli, R.E. and Machado, W. (2017). Metal Bioavailability in Contaminated Estuarine Sediments from a Highly-Impacted Tropical Bay. *Revista Virtual de Química*. [*Virtual Journal of Chemistry*] 9, 2007-2016.

Monte, C.N., Rodrigues, A.P.C., Freitas, A.R., Freire, A.S., Santelli, R. E., Braz, B.F. and Machado, W. (2019). Dredging impact on trace metal behavior in a polluted estuary: a discussion about sampling design. *Brazillian Journal of Oceanography*, 67, http://dx.doi.org/10.1590/S1679-87592019022706701

Monte, C.N., Rodrigues, A. P. C. Freitas, A.R., Braz, B.F., Freire, A. S., Cordeiro, R.C., Santelli, R. E., Machado, W. (2021). Ecological risks associated to trace metals of contaminated sediments from a densely urbanized tropical eutrophic estuary. *Environmental Monitoring and Assessment* 193(12)doi 10.1007/s10661-021-09552-7

Monteiro, F.F., Cordeiro, R.C., Santelli, R.E., Machado, W., Evangelista, H., Villar, L.S., Viana, L.C.A. and Bidone, E.D. (2012). Sedimentary geochemical record of historical anthropogenic activities affecting Guanabara Bay (Brazil) environmental quality. *Environmental Earth Sciences*, 65,1661–1669.

Montero, N.M.J., Belzunce-Segarra, M.J., Gonzalez, J-L., Menchaca, I., Garmendia, J.M., Etxebarria, N. and Franco, J. (2013). Application of Toxicity Identification Evaluation (TIE) procedures for the characterization and management of dredged harbor sediments. *Marine Pollution Bulletin, 71,* 259-268.

Morillo, J., Usero, J and Gracia, I. (2004). Heavy metal distribution in marine sediments from the southwest coast of Spain. *Chemosphere* 55, 431-442.https://doi.org/10.1016/j.chemosphere.2003.10.047.

Morse, J.W. (1994). Interactions of trace metals with authigenic sulfide minerals: implications for their bioavailability. *Marine Chemistry*, 46,.1-6.

Pazos-Capeáns, P., Barciela-Alonso, M.C., Herbello-Hermelo, P. and Bermejo-Barrera, P. (2010). Estuarine increase of chromium surface sediments: distribution, transport and time evolution. *Microchemical. Journal*, 96, 362–370.

Neyestani, M.R., Bastami, K.D., Esmaeilzadeh, M., Shemirani, F., Kazaali, A., Molamohyeddin, N., Afkham, M., Nourbakhsh, S., Dehghani, M., Aghaei, S. and Firouzbakht, M. (2016). Geochemical speciation and ecological risk assessment of selected metals in the surface sediments of the northern Persian Gulf. *Marine Pollution Bulletin*,15, 603-611.

Peña-Icart, M., Mendiguchia, C., Villanueva-Tagle, M.E., Pomares-Alfonso, M.S. and Moreno, C. (2014).Revisiting methods for the determination of bioavailable metals in coastal sediments. *Marine Pollution Bulletin*, 89, 67-74.

Pereira, E., Baptista-Neto, J.A., Smith, B.J. and McAllister, J. (2007). The contribution of heavy metal pollution derived from highway runoff to Guanabara Bay sediments – Rio de Janeiro/Brazil. *Annals of the Brazilian Academy of Sciences*, 79 (4), 739-750.

Rebello, A.L., Haekel, W., Moreira I., Santelli, R.E. and Schroeder, F. (1986).The fate of heavy metals in an estuarine tropical system. *Marine Chemistry,* 18,215–225.

Rickards, D. (2012). Metal sequestration by sedimentary iron sulfides. In: *Sulfid sediment and sedimentary rocks. Developments in Sedimentology*, 65, Chapter 7, 287-312.

Roberts, D.A. (2012). Causes and ecological effects of resuspended contaminated sediments (RCS) in marine environments. *Environment International*, 40:230–243.

Rodrigues, S.K. (2013). *Avaliação da disponibilidade potencial e toxicidade de metais-traço em sedimentos superficiais da região da Ilha da Madeira, Baía de Sepetiba,* [*Evaluation of the potential availability and toxicity of trace metals in surface sediments in the Madeira Island region, Sepetiba Bay*], RJ. Thesis, Master in Geochemistry, Universidade Federal Fluminense. 123p. In Portuguese.

Rodrigues, A.P.C., Lemos, A.P., Monte, C.N., Rodrigues, S.K., Cesar, R.G. and Machado, W. (2017). Environmental risk in a coastal zone from Rio de Janeiro state (Brazil) due to dredging activities. In: Araújo, C.V.M. & Shinn, C. *Ecotoxicology in Latin America.* Nova Publishers, New York, pp. 183-200.

Schintu, M., Marrucci, A., Marras, B., Galgani, F., Buosi, C., Ibba, A. and Cherchi, A. (2016). Heavy metal accumulation in surface sediments at the port of Cagliari (Sardinia, western Mediterranean): environmental assessment using sequential extractions and benthic foraminifera. *Marine Pollution Bulletin*,111, 45–56.

Shao, S., Liu, H., Tai, X., Zeng, F., Li, J. and Li, Y. (2020). Speciation and migration of heavy metals in sediment cores of urban wetland: bioavailability and risks. *Environmental Science Pollution Research.* https://doi.org/10.1007/s11356-020-08719-y

Silva e Silva, R., Blanco, C.J.C., Cavalcante, I.C.S., Teixera, L.C.G.M., Fernandes, L.L., Pessoa, F.C.L. (2020). Relationship between water quality parameters and land use of a small Amazonian catchment. *Sustainable Water Resources Management*, 6:65 https://doi.org/10.1007/s40899-020-00421-8.

Simpson, S.L., Apte, S.C., Batley, G.E. (1998). Effect of short-term resuspension events on trace metal speciation in polluted anoxic. *Environment Science and Technology*, 32, 620-625.

Vandenberg, C and Rebello, A.H. (1986). Organic-copper interactions in Guanabara Bay, Brazil - an electrochemical study of copper complexation by dissolved organic material in a tropical bay. *Science Total Environment*, 58(1-2): 37-45.

Van Den Berg, G.A., Meijers, G.G.A., Van Der Heijdt, L.M. and Zwolsman, J.J.G. (2001). Dredging-related mobilization of trace metals: a case study in the Netherlands. *Water Research*, 35, 1979-1986.

Zhang, C., Zhi-Gang, Y.U., Zeng, Z., Jiang, G., Yang, M., Cui, F., Zhu, M., Shen, L and Hu, L. (2014) Effects of sediment geochemical properties on heavy metal bioavailability. *Environmental International*, 73, 270-281.

Wu, J, Li P., Qian, H., Duan, Z. and Zhan, X. (2014). Using correlation and multivariate statistical analysis to identify hydrogeochemical processes afectıng the major ion chemistry of waters: case study in Laoheba phosphorite mine in Sichuan, China. *Arabian Journal Geosciences* 7(10),3973–3982. https://doi.org/10.1007/s12517-013-1057-4.

Wu J, Li P., Wang, D., Ren, X. and Wei, M. (2019). Statistical and multivariate statistical techniques to trace the sources and afecting factors of groundwater pollution in a rapidly growing city on the Chinese Loess Plateau. *Human and Ecological Risk Assessment: An International Journal,* https://doi.org/10.1080/10807039.2019.1594156.

Xin, Z; Taofa, Z and Xifei, Y. (2005). Study on Assessment Methods of Heavy Metal Pollution in River Sediments. *Journal of Hefei University of Technology (Natural Science*), 28 (11): 1419- 1423.

Biographical Sketch

Christiane do Nascimento Monte

Affiliation: Federal University of Western of Pará.

Education: PhD Business Address: Rua Vera Paz SN, Salé, Santarém, Pará, Brazil

Research and Professional Experience: environmental geochemistry, biogeochemistry, water quality, environmental geology, sediment quality, trace metals, risk assessment, ecotoxcology.

Professional Appointments: Environmental Geology Professor in Federal University of Western of Pará, reviewer for the journal Water Research and other journals of environmental quality and management, and environmental researcher.

Publications from the Last 3 Years:

1. Do Nascimento monte, Christiane, Correa, Edinelson, Costa, Igor, Nascimento, Thiago, Pereira, Mateus, Batista, Louisiane, Pinheiro, Danilo. The physical-chemical characteristics of surface waters in the management of quality in clear water rivers in the Brazilian Amazon. *Water Policy,* v. 23, p. 1089, 2021.
2. Monte, C. N., Rodrigues, A. P. C., Macedo, S., Regis, C. R., Correa, Edinelson Saldanha, Ribeiro, A. C., Machado, W. T. V. A influência antrópica na qualidade da água do rio Tapajós, na cidade de Santarém-PA.[The anthropic influence on the water quality of the Tapajós River in the city of Santarém-PA.] *Revista Brasileira De Geografia Física* [*Brazilian Journal of Physical Geography*], v. 14, p. 3279, 2021.
3. Do Nascimento Monte, Christiane, De Castro Rodrigues, Ana Paula, De Freitas, Alexandre Rafael, Braz, Bernardo Ferreira, Freire, Aline Soares, Cordeiro, Renato Campello, Santelli, Ricardo Erthal, Machado, Wilson Thadeu Valle. Ecological risks associated to trace metals of contaminated sediments from a densely urbanized tropical eutrophic estuary. *Environmental Monitoring and Assessment,* v. 193, p. 193, 2021.
4. Correa, Edinelson Saldanha, Monte, Christiane, Nascimento, Thiago Shinaigger Rocha Do. Avaliação de impacto ambiental causado por efluentes da estação de piscicultura Santa Rosa, Santarém/Pará. [Environmental impact assessment caused by effluents from the Santa Rosa fish farm station, Santarém/Pará]

Revista Ibero-americana de Ciências Ambientais [*Ibero-American Journal of Environmental Sciences*], v. 11, p. 260-273, 2020.

5. Ferreira, A. R. L., Cesar, R. G., Siqueira, D.M, Rodrigues, A. P. C., Vezzone, M., Monte, C. N., Machado, Wilson, Castilhos, Z, Campos, T, Polivanov, H, Leite, S. G. F. Potencial Tóxico De Sedimentos Dragados Das Baías De Sepetiba E Da Guanabara (Rj) Em Cenário De Disposição Em Latossolo. [Toxic Potential Of Dredged Sediments From Sepetiba Bays And Guanabara (Rj) In Latosol Disposition Scenario.] *Geociências* [*Geosciences*] (São Paulo. Online), v. 39, p. 1141, 2020.
6. Costa, Igor, Saldanha, Edinelson Correa, Monte, Christiane Do Nascimento. A sazonalidade de contaminantes em águas subterrâneas e superficiais entorno de um aterro sanitário na região Amazônica. [The seasonality of contaminants in groundwater and surface water around a landfill in the Amazon region.] *Revista Ibero-americana de Ciências Ambientais* [*Ibero-American Journal of Environmental Sciences*], v. 11, p. 371-382, 2020.
7. Santos, R. S., Monte, C. N.. Risco geológico decorrente da erosão hídrica na zona urbana de Santarém- ParáBrasil. Caso de estudo: avenida Bugaville. [Geological risk resulting from water erosion in the urban area of Santarém- ParáBrasil. Case study: Bugaville Avenue.] *Brazilian Journal of Animal and Environmental Research,* v. 2, p. 526, 2019.
8. Vale, R. S., Lima, L. S., Monte, C. N., Santana, R. A. S.. Evidências do fenômeno de terras caídas com grandes cheias na região Oeste do Pará. [Evidence of the phenomenon of fallen lands with large floods in the Western region of Pará.] *Brazilian Journal of Development,* v. v. 5, p. 6295-6302, 2019.
9. Monte, Christiane Do Nascimento, Rodrigues, Ana Paula De Castro, De-Freitas, Alexandre Rafael, Freire, Aline Soares, Santelli, Ricardo Erthal, Braz, Bernardo Ferreira, Machado, Wilson. Dredging impact on trace metal behavior in a polluted estuary: a discussion about sampling design. *Brazilian Journal of Oceanography (Online),* v. 67, p. 1, 2019.
10. Campos, Bruno Galvão De, Moreira, Lucas Buruaem, Pauly, Guacira De Figueiredo Eufrasio, Cruz, Ana Carolina Feitosa, Monte, Christiane Do Nascimento, Dias Da Silva, Lílian Irene, Rodrigues, Ana Paula De Castro, Machado, Wilson, Abessa, Denis

Moledo De Souza. Integrating multiple lines of evidence of sediment quality in a tropical bay (Guanabara Bay, Brazil). *Marine Pollution Bulletin,* v. 146, p. 925-934, 2019.

11. Freitas, A.R., Rodrigues, Ana Paula De Castro, Monte, C. N., Freire, Aline Soares, Santelli, Ricardo Erthal, Machado, Wilson, Sabadini-Santos, Elisamara. Increase in the bioavailability of trace metals after sediment resuspension. *SN Applied Sciences,* v. 1, p. 1288, 2019.
12. Monte, Christiane Do Nascimento, Saldanha, Edinelson Correa, Pinheiro, Danilo Costa. Índice de estado trófico e a proveniência do fósforo e clorofila-a em diferentes estações do ano em uma microbacia Amazônica. [Trophic state index and the provenance of phosphorus and chlorophyll-a in different seasons in an Amazon basin.] *Revista Ibero-Americana de Ciências Ambientais* [*Ibero-American Journal of Environmental Sciences*], v. 10, p. 89-100, 2019.
13. Vezzone, M., Cesar, R. G., Polivanov, H., Serrano, A. F., Nascimento, M. T., Siqueira, D. M., Rodrigues, A. P. C., Monte, C., Castilhos, Z. C., Campos, T. M. P.. Influence of Salinity on the Toxicity of Dredged Sediments from Estuarine Rodrigo de Freitas Lagoon and Guanabara Bay (RJ): Toxic Effects on Earthworms. *Anuário do Instituto de Geociências* (Ufrj. Impresso) [*Yearbook of the Institute of Geosciences* (Ufrj. Printed)], v. 42, p. 07-17, 2019.
14. Monte, Christiane, Cesar, Ricardo, Rodrigues, Ana Paula, Siqueira, Danielle, Serrano, Aline, Abreu, Leticia, Teixeira, Matheus, Vezzone, Mariana, Polivanov, Helena, Castilhos, Zuleica, De Campos, Tácio, Machado, Glaucia G. M., Souza, Weber F., Machado, Wilson. Spatial variability and seasonal toxicity of dredged sediments from Guanabara Bay (Rio de Janeiro, Brazil): acute effects on earthworms. *Environmental Science and Pollution Research,* v. 25, p. 1, 2018.

Bibliography

Behaviors of Trace Metals in Environment: The Pollution in Regional and Metropolis Areas

LCCN	2019762294
Type of material	Book
Personal name	Zhang, Hui, author.
Main title	Behaviors of Trace Metals in Environment: The Pollution in Regional and Metropolis Areas / by Hui Zhang.
Edition	1st ed. 2020.
Published/Produced	Singapore: Springer Singapore: Imprint: Springer, 2020.
Description	1 online resource (XII, 364 pages 147 illustrations, 25 illustrations in color.) text file PDF
ISBN	9789811336126
Summary	This book focuses on the behavior and impact of trace metals in the environment by studying typical cases from China such as the Hetao Area of the Yellow River, Shanghai, and Nanjing. Based on samples and experiments on the behavior of pollutants, it systematically discusses the regulation of trace metals' distribution, accumulation, and migration, associated with the cause of formation demonstration. The author subsequently uses the acquired data to review the evolving trend of trace metal behaviors in natural systems (river or lake water, sediments, and soils), develops suggestions for the prevention of their negative effects, and devise treatments. Moreover, he proposes solutions to difficult research issues such as trace metal speciation extraction, and an analysis, along with operational procedures. Given its scope, the book will provide a valuable guide for researchers and

	engineers in relevant disciplines of the environmental sciences and engineering, and for environmental policymakers to consult in practices.
Contents	The Regional Pollution of Trace Metals-the Hetao Area, China -- The Metropolis Pollution of Trace Metals - Shanghai and Nanjing, China -- The Experimental Research on the Behaviors of Trace Metals -- An Approach on the Behavior Impacts and Factors of Trace Metals in Environment -- The Speciation of Trace Metals and Research Methods -- An Approach to the Identification for the Original and Added Concentrations of Trace Metals in Soil System Polluted by Trace Metals -- Main Research Results.
Subjects	Environmental chemistry. Environmental geology. Environmental monitoring. Geoecology. Environmental Chemistry. Geoecology/Natural Processes. Monitoring/Environmental Analysis.
Notes	Description based on publisher-supplied MARC data.
Additional formats	Print version: Behaviors of trace metals in environment. 9789811336119 (DLC) 2019931535 Printed edition: 9789811336119 Printed edition: 9789811336133

Behaviors of trace metals in environment.

LCCN	2019931535
Type of material	Book
Main title	Behaviors of trace metals in environment.
Published/Produced	New York, NY: Springer Berlin Heidelberg, 2019.
ISBN	9789811336119

Biogas Science and Technology

LCCN	2019756457
Type of material	Book

Main title	Biogas Science and Technology / edited by Georg Gübitz, Alexander Bauer, Guenther Bochmann, Andreas Gronauer, Stefan Weiss.
Edition	1st ed. 2015.
Published/Produced	Cham: Springer International Publishing: Imprint: Springer, 2015.
Description	1 online resource (V, 200 pages 23 illustrations, 8 illustrations in color.) text file PDF
ISBN	9783319219936
Related names	Bauer, Alexander. editor. Bochmann, Guenther. editor. Gronauer, Andreas. editor. Gübitz, Georg. editor. Weiss, Stefan. editor.
Summary	Michael Lebuhn, Stefan Weiss, Bernhard Munk, Georg M. Guebitz Microbiology and Molecular Biology Tools for Biogas Process Analysis, Diagnosis and Control Veronika Dollhofer, Sabine Marie Podmirseg, Tony Martin Callaghan, Gareth Wyn Griffith and Katerina Fliegerová Anaerobic Fungi and their Potential for Biogas Production Bianca Fröschle, Monika Heiermann, Michael Lebuhn, Ute Messelhäusser, Matthias Plöchl Hygiene and Sanitation in Biogas Plants Charles-David Dubé and Serge R. Guiot Direct Interspecies Electron Transfer in Anaerobic Digestion: A Review Simon K.-M. R. Rittmann A Critical Assessment of Microbiological Biogas to Biomethane Upgrading Systems Manfred Lübken, Pascal Kosse, Konrad Koch, Tito Gehring, Marc Wichern Influent Fractionation for Modeling Continuous Anaerobic Digestion Processes Fermoso, F. G, van Hullebusch, E. D, Guibaud, G, Collins, G, Svensson, B. H, Carliell-Marquet, C, Vink, J.P.M, Esposito, G, Frunzo, L Fate of Trace Metals in Anaerobic Digestion.
Subjects	Biotechnology.

	Biotechnology.
Notes	Description based on publisher-supplied MARC data.
Additional formats	Print version: Biogas science and technology. 9783319219929 (DLC) 2015945326 Printed edition: 9783319219929 Printed edition: 9783319219943 Printed edition: 9783319358802
Series	Advances in Biochemical Engineering/ Biotechnology, 0724-6145; 151 Advances in Biochemical Engineering/ Biotechnology, 0724-6145; 151

Biogeochemistry of trace elements

LCCN	2018418632
Type of material	Book
Main title	Biogeochemistry of trace elements / Oleg S. Pokrovsky and Jerome Viers, editors.
Published/Produced	New York: Nova Science Publishers, Inc., [2018]
Description	x, 384 pages: illustrations, maps; 26 cm.
ISBN	1536142441 hardcover 9781536142440 hardcover ebook
LC classification	QH343.7 .B565 2018
Related names	Pokrovsky, Oleg S., editor. Viers, Jerome, editor.
Contents	Introduction -- Trace Metal Exposure in Different Livestock Production Systems / I. Orjales, R. Rodríguez-Bermúdez, M. Miranda, M. López-Alonso and M. Garcia-Vaquero -- Lithological Distribution of Rare Earth Elements in Soil and Atmospheric Precipitates in the Bregalnica River Basin / Trajče Stafilov, Biljana Balabanova and Robert Šajn -- Identification, Evaluation, and Estimation of the Levels of Potentially Harmful Trace Elements in Sediments Based on the Application of Different Methods / Sanja M. Sakan, Nenad M. Sakan and Dragana S. Đorđević --

Accumulation of Trace Elements in Sediments and Macrophytes of Thermokarst Lakes in Western Siberia / R.M. Manasypov, O.S. Pokrovsky, L.S. Shirokova, S.N. Kirpotin and N.S. Zinner -- Trace Elements in Snow Cover of Western Siberia: Impact of Snow Deposition on Surface Water Chemistry / V.P. Shevchenko, S.N. Vorobyev, I.V. Krickov, R.M. Manasypov, N.V. Politova, S.G. Kopysov, O.M. Dara, Y. Auda, L.S. Shirokova, L.G. Kolesnichenko, V.A. Zemtsov, S.N. Kirpotin and O.S. Pokrovsky -- Major and Trace Elements in Peat Pore Water Found in the Permafrost Zone of Western Siberia / T.V. Raudina, S.V. Loiko, A. Lim, I.V. Krickov, L.S. Shirokova, G.I. Istigechev, D.M. Kuzmina, S.P. Kulizhsky, S.N. Vorobyev and O.S. Pokrovsky -- Hot Spots of Permafrost Thawing Enhance Trace Metal Release into Thermokarst Waters / S.V. Loiko, O.S. Pokrovsky, T.V. Raudina, A.G. Lim, L.G. Kolesnichenko, L.S. Shirokova, S.N. Vorobyev, S.N. Kirpotin and D.M. Kuzmina -- Trace Elements in the Form of Organo-Mineral Colloids in the Mixing Zone of the Arctic River / O.S. Pokrovsky, J. Viers, A.V. Chupakov, L.S. Shirokova, V.V. Gordeev and V.P. Shevchenko -- The Distribution of Metals in Different Types of Soils of Northern Karelia / O. Yu. Drozdova, Yu. A. Zavgorodnyaya, D.A. Bychkov, V.V. Demin and S.A. Lapitskiy -- Trace Metals in Soil Catenas of the Arctic Islands (THe Svalbard and Novaya Zemlya Archipelagos) / Vidas V. Kriauciunas, Stanislav A. Iglovsky and Irina A. Kuznetsova -- Trace Metal Binding Properties of the Peat: Static and Dynamic Sorption / Irina A. Kuznetsova, Stanislav A. Igkovsky and Vidas V. Kriauciunas

Subjects Biogeochemistry.
Trace elements.
Biogeochemistry.

Notes Includes bibliographical references and index.

Series	Chemistry research and applications Chemistry research and applications series.

Biogeochemistry of trace elements

LCCN	2020685631
Type of material	Book
Main title	Biogeochemistry of trace elements / Oleg S. Pokrovsky and Jerome Viers, editors.
Published/Produced	New York: Nova Science Publishers, Incorporated, [2018]
Description	1 online resource
ISBN	9781536142457 ebook hardcover hardcover
LC classification	QH343.7
Related names	Pokrovsky, Oleg S., editor. Viers, Jerome, editor.
Contents	Introduction -- Trace Metal Exposure in Different Livestock Production Systems / I. Orjales, R. Rodríguez-Bermúdez, M. Miranda, M. López-Alonso and M. Garcia-Vaquero -- Lithological Distribution of Rare Earth Elements in Soil and Atmospheric Precipitates in the Bregalnica River Basin / Trajče Stafilov, Biljana Balabanova and Robert Šajn -- Identification, Evaluation, and Estimation of the Levels of Potentially Harmful Trace Elements in Sediments Based on the Application of Different Methods / Sanja M. Sakan, Nenad M. Sakan and Dragana S. Đorđević -- Accumulation of Trace Elements in Sediments and Macrophytes of Thermokarst Lakes in Western Siberia / R.M. Manasypov, O.S. Pokrovsky, L.S. Shirokova, S.N. Kirpotin and N.S. Zinner -- Trace Elements in Snow Cover of Western Siberia: Impact of Snow Deposition on Surface Water Chemistry / V.P. Shevchenko, S.N. Vorobyev, I.V. Krickov, R.M. Manasypov, N.V. Politova, S.G. Kopysov, O.M. Dara, Y. Auda, L.S. Shirokova,

L.G. Kolesnichenko, V.A. Zemtsov, S.N. Kirpotin and O.S. Pokrovsky -- Major and Trace Elements in Peat Pore Water Found in the Permafrost Zone of Western Siberia / T.V. Raudina, S.V. Loiko, A. Lim, I.V. Krickov, L.S. Shirokova, G.I. Istigechev, D.M. Kuzmina, S.P. Kulizhsky, S.N. Vorobyev and O.S. Pokrovsky -- Hot Spots of Permafrost Thawing Enhance Trace Metal Release into Thermokarst Waters / S.V. Loiko, O.S. Pokrovsky, T.V. Raudina, A.G. Lim, L.G. Kolesnichenko, L.S. Shirokova, S.N. Vorobyev, S.N. Kirpotin and D.M. Kuzmina -- Trace Elements in the Form of Organo-Mineral Colloids in the Mixing Zone of the Arctic River / O.S. Pokrovsky, J. Viers, A.V. Chupakov, L.S. Shirokova, V.V. Gordeev and V.P. Shevchenko -- The Distribution of Metals in Different Types of Soils of Northern Karelia / O. Yu. Drozdova, Yu. A. Zavgorodnyaya, D.A. Bychkov, V.V. Demin and S.A. Lapitskiy -- Trace Metals in Soil Catenas of the Arctic Islands (THe Svalbard and Novaya Zemlya Archipelagos) / Vidas V. Kriauciunas, Stanislav A. Iglovsky and Irina A. Kuznetsova -- Trace Metal Binding Properties of the Peat: Static and Dynamic Sorption / Irina A. Kuznetsova, Stanislav A. Igkovsky and Vidas V. Kriauciunas.

Subjects	Biogeochemistry. Trace elements. Biogeochemistry.
Notes	Includes bibliographical references and index. Description based on print version record.
Additional formats	Print version: Biogeochemistry of trace elements New York: Nova Science Publishers, Inc., [2018] 1536142441 (DLC) 2018418632
Series	Chemistry research and applications Chemistry research and applications series.

Commercial Surfactants for Remediation: Mobilization of Trace Metals from Estuarine Sediment and Bioavailability

LCCN	2019763024
Type of material	Book
Personal name	Bisht, Anu Singh, author.
Main title	Commercial Surfactants for Remediation: Mobilization of Trace Metals from Estuarine Sediment and Bioavailability / by Anu Singh Bisht.
Edition	1st ed. 2019.
Published/Produced	Singapore: Springer Singapore: Imprint: Springer, 2019.
Description	1 online resource (XIX, 100 pages 36 illustrations) text file PDF
ISBN	9789811302213
Summary	This book demonstrates the benefits of using commercially available surfactants, or surface-active agents, for remediation of metal-contaminated soil and sediment. First the book offers theoretical reviews of commercially available surfactants, then it proceeds to a study of various available surfactants for the mobilization of metals. Surfactants representative of amphiphiles discovered in the digestive environment of sediment-ingesting organisms are used to examine the extent and rate of metal (Al, Fe, Cd, Cu, Mn, Ni, Pb, Sn, Zn) mobilization from contaminated estuarine sediment. Metals can cause harmful effects to the environment and organisms. It is difficult to treat contaminants that are often tightly bound to the extremely small size of the estuarine sediments. The book also demonstrates the mechanisms of metal mobilization that appear to be related to complexation with monomers and adsorption to micelles of the anionic amphiphiles, and to the denudation of hydrophobic host phases or coatings on the sediment by micelles of both anionic and nonionic surfactants. Readers obtain a better understanding of current commercial surfactants,

their impact on the environment, and possible remediation. This transdisciplinary book contributes toward Sustainable Development Goals numbers 6 (Clean Water and Sanitation) and 13 (Climate Action) set by the United Nations and is useful for students and teachers of sediment studies, coastal studies, environmental sciences, hydrology, civil engineering, and policy sciences.

Contents Introduction -- Bioavailability of Metals in Sediment -- Solubilization of Metals in Invertebrate Guts -- Surfactants -- Commercial Surfactants for Remediation- Introduction -- Commercial Surfactants for Remediation- Methodology -- Metal Analysis -- Metal Concentration -- Kinetics of Metal Release by Surfactants -- Surfactant Availability of Metals -- Implication of Surfactants in Remediation -- General Discussion and Conclusions -- Marine Conservation and Sustainable Development Goals.

Subjects Coasts.
Environmental chemistry.
Environmental sciences.
Soil conservation.
Soil science.
Sustainable development.
Environmental Science and Engineering.
Coastal Sciences.
Environmental Chemistry.
Soil Science & Conservation.
Sustainable Development.

Notes Description based on publisher-supplied MARC data.

Additional formats Print version: Surfactants for environmental remediation. 9789811302206 (DLC) 2018955718
Printed edition: 9789811302206
Printed edition: 9789811302220
Printed edition: 9789811343636

Series	Advances in Geographical and Environmental Sciences, 2198-3542 Advances in Geographical and Environmental Sciences, 2198-3542

Environmental Contaminants: Using natural archives to track sources and long-term trends of pollution

LCCN	2019764703
Type of material	Book
Main title	Environmental Contaminants: Using natural archives to track sources and long-term trends of pollution / edited by Jules M. Blais, Michael R. Rosen, John P. Smol.
Edition	1st ed. 2015.
Published/Produced	Dordrecht: Springer Netherlands: Imprint: Springer, 2015.
Description	1 online resource (XVI, 509 pages 102 illustrations, 44 illustrations in color.) text file PDF
ISBN	9789401795418
Related names	Blais, Jules M, editor. Rosen, Michael R, editor. Smol, John P, editor.
Summary	The human footprint on the global environment now touches every corner of the world. This book explores the myriad ways that environmental archives can be used to study the distribution and long-term trajectories of chemical contaminants. The volume first focuses on reviews that examine the integrity of the historic record, including factors related to hydrology, post-depositional diffusion, and mixing processes. This is followed by a series of chapters dealing with the diverse archives and methodologies available for long-term studies of environmental pollution, such as the use of sediments, ice cores, sclerochronology, and museum specimens.

Contents	1. Using natural archives to track sources and long-term trends of pollution -- 2. The influence of hydrology on lacustrine sediment contaminant records -- 3. The stability of metal profiles in freshwater and marine sediments -- 4. Calculating rates and dates and interpreting contaminant profiles in biomixed sediments -- 5. Contaminants in marine sedimentary deposits from coal fly ash during the Latest Permian Extinction -- 6. Lake sediment records of preindustrial metal pollution. Colin Cooke and Richard Bindler -- 7. Lacustrine archives of metals from mining and other industrial activities -- 8. Organic pollutants in sediment core archives -- 9. Environmental archives of contaminant particles -- 10. Long range atmospheric transport in Arctic regions using lake sediments -- 11. Tracking long-range atmospheric transport of trace metals, polycyclic aromatic hydrocarbons, and organohalogen compounds using lake sediments of mountain regions -- 12. Tracking contaminant transport from biovectors -- 13. Using peat records as natural archives of past atmospheric metal deposition -- 14. Historical contaminant records from sclerochronological archives -- 15. Contaminant records in ice cores -- 16. Use of catalogued long-term biological collections and samples for determining changes in contaminant exposure to organisms.-Chapter 17. Using natural archives to track sources and long-term trends of pollution: Some final thoughts and suggestions for future directions.
Subjects	Ecotoxicology. Environmental chemistry. Environmental management. Hydrology. Water pollution. Water quality.

	Waste Water Technology / Water Pollution Control / Water Management / Aquatic Pollution. Ecotoxicology. Environmental Chemistry. Hydrology/Water Resources. Water Policy/Water Governance/Water Management. Water Quality/Water Pollution.
Notes	Description based on publisher-supplied MARC data.
Additional formats	Print version: Environmental contaminants. 9789401795401 (DLC) 2014955353 Printed edition: 9789401795401 Printed edition: 9789401795425 Printed edition: 9789402403664
Series	Developments in Paleoenvironmental Research, 1571-5299; 18 Developments in Paleoenvironmental Research, 1571-5299; 18

Environmental Sustainability: Role of Green Technologies

LCCN	2019750276
Type of material	Book
Main title	Environmental Sustainability: Role of Green Technologies / edited by P. Thangavel, G. Sridevi.
Edition	1st ed. 2015.
Published/Produced	New Delhi: Springer India: Imprint: Springer, 2015.
Description	1 online resource (XVII, 324 pages 58 illustrations, 32 illustrations in color.) text file PDF
ISBN	9788132220565
Related names	Sridevi, G. editor. Thangavel, P. editor.
Summary	Covers different categories of green technologies (e.g., biofuels, renewable energy sources, phytoremediation et cetera,) in a nutshell - Focuses on next generation technologies which will help to attain the sustainable development - The chapters

widely cover for students, faculties and researchers in the scientific arena of Environmentalists, Agriculturalists, Engineers and Policy Makers The World Environment Day 2012 is prepared to embrace green economy. The theme for 2012 encompasses various aspects of human living, ranging from transport to energy to food to sustainable livelihood. Green technology, an eco-friendly clean technology contributes to sustainable development to conserve the natural resources and environment which will meet the demands of the present and future generations. The proposed book mainly focuses on renewable energy sources, organic farming practices, phyto/bioremediation of contaminants, biofuels, green buildings and green chemistry. All of these eco-friendly technologies will help to reduce the amount of waste and pollution and enhance the nation's economic growth in a sustainable manner. This book is aimed to provide an integrated approach to sustainable environment and it will be of interest not only to environmentalists but also to agriculturists, soil scientists and bridge the gap between the scientists and policy-makers..

Contents 1. Insight into the Role of Arbuscular Mycorrhizal Fungi in Sustainable Agriculture -- 2. Recycled Water Irrigation for Sustainable Production in Australia -- 3. A Review of Biopesticides and their Mode of Action against Insect Pests -- 4. Seaweeds-a Promising Source for Sustainable Development -- 5. A Comprehensive Overview of Renewable Energy Status in India -- 6. How Sahara Solar Breeder Plan Contribute to Global Sustainable Energy Production, Si Manufacturing, Resource Conservation and help Balance Societies Needs? -- 7. Clean Development Mechanism: A key to Sustainable Development -- 8. Microalgae as an Attractive Source for Biofuel Production -- 9.

Advancement and Challenges in Harvesting Techniques for Recovery of Microalgae Biomass -- 10. Characterization of Bacillus Strains Producing Biosurfactants -- 11. Production of Biosurfactants using Eco-friendly Microorganisms -- 12. Ecofriendly Technologies for Heavy Metal Remediation - A Pragmatic Approaches -- 13. Phytoextraction of Trace Metals - Principles and Applications -- 14. Integrated Management of Mine Waste using Biogeotechnologies Focusing Thai Mines -- 15. Constructed Wetlands: An Ecotechnology for Wastewater Treatment and Conservation of Ganga Water Quality -- 16. Mycorrhizal Plants Accelerated Revegetation on Coal Mine Overburden in the Dry Steppes of Kazakhstan -- 17. Drivers of Green Economy: An Indian Perspective -- 18. Green Nanotechnology: The Solution to Sustainable Development of Environment.

Subjects Biotechnology.
Ecology.
Environmental engineering.
Renewable energy sources.
Sustainable development.
Sustainable Development.
Ecology.
Environmental Engineering/Biotechnology.
Renewable and Green Energy.

Notes Description based on publisher-supplied MARC data.

Additional formats Print version: Environmental sustainability 9788132220558 (DLC) 2014952888
Printed edition: 9788132220558
Printed edition: 9788132220572
Printed edition: 9788132229520

Management of Natural Resources in a Changing Environment

LCCN 2019766683

Type of material	Book
Main title	Management of Natural Resources in a Changing Environment / edited by N. Janardhana Raju, Wolfgang Gossel, M. Sudhakar.
Edition	1st ed. 2015.
Published/Produced	Cham: Springer International Publishing: Imprint: Springer, 2015.
Description	1 online resource (XVI, 298 pages 100 illustrations, 75 illustrations in color.) text file PDF
ISBN	9783319125596
Related names	Gossel, Wolfgang, editor. Raju, N. Janardhana, editor. Sudhakar, M, editor.
Summary	This book addresses issues related to sources of groundwater pollution such as arsenic, uranium, fluoride and their effects on human health. It discusses extensively the removal of heavy metals, arsenic and fluoride from drinking water. Bioremediation and phyto remediation on biomass productivity are treated in several chapters in the book. The volume highlights leachate characteristics analysed both in the laboratory and in field studies assessing the trace metals in rainwater. This book is a study on the judicious management of natural resources and exposes environmental problems particularly those related to pollution and bioremediation.
Contents	Message from Alexander von Humboldt Foundation -- Foreword -- Preface -- About the Editors -- Hydro-geochemical Investigation and Quality Assessment of Groundwater for Drinking and Agriculture Use in Jawaharlal Nehru University (JNU), New Delhi, India - Comparison of Relationship Between the Concentrations of Water Isotopes in Precipitation in the Cities of Tehran (Iran) and New Delhi (India) -- Geophysical Expression for Groundwater Quality in Part of

Chittoor District, Andhra Pradesh, India -- Geospatial Analysis of Fluoride Contamination in Groundwater of Southeastern Part of Anantapur District, Andhra Pradesh -- Identification of Surface Water Harvesting Sites for Water Stressed Area Using GIS: A Case Study of Ausgram Block, BurdwanDistrict, West Bengal, India -- Forecasting Groundwater Level Using Hybrid Modelling Technique -- Alterations in Physico-chemical Parameters of Water and Aquatic Diversity at Maneri-Bhali Phase I Dam Site on River Ganges in District Uttarkashi, Uttarakhand -- Effective Removal of Heavy Metals and Dyes from Drinking Water Utilizing Bio-compatible Magnetic Nanoparticle -- UASBR: An Effective Wastewater Treatment Option to Curb Greenhouse Gas Emissions -- Biogas Upgrading and Bottling Technology for Vehicular and Cooking Applications -- Use of Indigenous Bacteria from Arsenic Contaminated Soil for Arsenic Bioremediation -- Adsorption of Arsenite and Fluoride on Untreated and Treated Bamboo Dust -- Reducing the Toxicity of Carbon Nanotubes and Fullerenes Using Surface Modification Strategy -- Phytoremediation Study and Effect of pH on Biomass Productivity of Eichhornia crassipes -- Regeneration of White Oak (Quercus leucotrichophora) in Two Pine Invaded Forests in Indian Central Himalaya -- Human Health Risk Assessment of Heavy Metals from Bhalaswa Landfill, New Delhi, India -- Transport of Trace Metals by the Rainwater Runoff in the Urban Catchment of Guwahati, India -- Analysis of Leachate Characteristics to Study Coal Ash Usability -- Air Pollution Mapping and Quality Assessment Study at an Urban Area Tirupati Using GIS -- Environmental Hazards and Conservation Approach to the Biodiversity and Ecosystem of the

St. Martin's Island in Bangladesh -- Uranium Toxicity in the State of Punjab in North-Western India -- Fluoride Toxicity in the Fluoride Endemic Villages of Gaya District, Bihar, India -- Index.

Subjects Biodiversity.
Environmental geology.
Environmental management.
Environmental pollution.
Geoecology.
Waste management.
Geoecology/Natural Processes.
Biodiversity.
Environmental Management.
Terrestrial Pollution.
Waste Management/Waste Technology.

Notes Description based on publisher-supplied MARC data.

Additional formats Print version: Management of natural resources in a changing environment 9783319125589 (DLC) 2014957126
Printed edition: 9783319125589
Printed edition: 9783319125602

Marine geochemistry: ocean circulation, carbon cycle and climate change

LCCN 2016932193

Type of material Book

Personal name Roy-Barman, Matthieu, author.

Main title Marine geochemistry: ocean circulation, carbon cycle and climate change / Matthieu Roy-Barman and Catherine Jeandel.

Edition First edition.

Published/Produced Oxford: Oxford University Press, 2016.

Description xxiii, 398 pages: illustrations (some color); 25 cm

ISBN 0198787499
9780198787495
9780198787501
0198787502

LC classification	GC111.2 .R685 2016
Related names	Jeandel, Catherine, author.
Summary	"Marine geochemistry uses chemical elements and their isotopes to study how the ocean works in terms of ocean circulation, chemical composition, biological activity and atmospheric CO2 regulation. This rapidly growing field is at a crossroad for many disciplines (physical, chemical and biological oceanography, geology, climatology, ecology, etc.). It provides important quantitative answers to questions such as: What is the deep ocean mixing rate? How much atmospheric CO2 is pumped by the ocean? How fast are pollutants removed from the ocean? How do ecosystems react to anthropogenic pressure? This text gives a simple introduction to the concepts, the methods and the applications of marine geochemistry with a particular emphasis on isotopic tracers. Overall introducing a very large number of topics (physical oceanography, ocean chemistry, isotopes, gas exchange, modelling, biogeochemical cycles), with a balance of didactic and indepth information, it provides an outline and a complete course in marine geochemistry. Throughout, the book uses a hands-on approach with worked out exercises and problems (with answers provided at the end of the book), to help the students work through the concepts presented. A broad scale approach is take including ocean physics, marine biology, ocean-climate relations, remote sensing, pollutions and ecology, so that the reader acquires a global perspective of the ocean. It also includes new topics arising from ongoing research programs. This textbook is essential reading for students, scholars, researchers and other professionals" -- Provided by publisher.
Contents	Machine generated contents note: 1. A Few Bases of Descriptive and Physical Oceanography -- 1.1. The Size of the Ocean -- 1.2. Salinity, Temperature

and Density: The Basic Parameters of the Oceanographer -- 1.2.1. Salinity -- 1.2.2. Temperature -- 1.2.3. Density -- 1.3. Vertical Structure of the Ocean -- 1.4. The Main Water Masses -- 1.5. Ocean Currents -- 1.5.1. Surface Circulation -- 1.5.2. The Physical Principles -- 1.5.3. The Wind-Driven Ocean Circulation -- 1.5.4. Ekman Pumping -- 1.5.5. Coastal Upwelling -- 1.5.6. Geostrophic Currents -- 1.6. Large-Scale Circulation -- 1.6.1. Vorticity -- 1.6.2. Sverdrup Balance -- 1.6.3. The Intensification of the Western Boundary Currents -- 1.6.4. Eddies and Recirculation -- 1.6.5. The Thermocline Ventilation -- 1.6.6. The Equatorial Circulation -- 1.6.7. The Deep Circulation -- Appendix 1. The Atmospheric Forcing -- Problems -- 2. Seawater Is More than Salted Water -- 2.1. Why Is Seawater Salty? -- 2.1.1. The Chemical Composition of Salt – 2.1.2. Residence Time -- 2.1.3. Rivers and Estuaries -- 2.1.4. The Atmosphere -- 2.1.5. Volcanic and Hydrothermal Processes -- 2.1.6. The Removal of Chemical Elements -- 2.2. Concept of Conservative and Non-Conservative Tracers -- 2.3. The Nutrient Cycle and the Role of Biological Activity -- 2.3.1. Nutrient Profiles in Seawater -- 2.3.2. The Life Cycles in the Ocean -- 2.3.3. Influence of Deep Circulation on the Nutrient Distribution -- 2.4. Gases in Seawater -- 2.4.1. Definition of Apparent Oxygen Utilization -- 2.5. Relationships between the Different Tracers -- 2.5.1. Extracting the Conservative Fraction of a Tracer -- 2.5.2. Construction of Conservative Tracers -- 2.5.3. Horizontal and Vertical Changes of Tracers -- 2.6. Carbon Chemistry -- 2.6.1. The Carbonate System -- 2.6.2. Calcium Carbonate -- 2.6.3. Organic Carbon -- 2.7. The Redox Conditions in the Ocean -- 2.8. Behavior of Trace Metals -- 2.8.1. The Different Types of Profiles -- 3.10. Silicon Isotope

Fractionation -- 3.11. Iron Isotope Fractionation -- 3.12. Mixing of Isotopic Tracers -- 3.12.1. Conservative Mixing -- 3.12.2. Non-Conservative Mixing -- 3.13. Evolution of the Isotopic Signature during a Reaction -- 3.13.1. Example: Nitrate Assimilation by Phytoplankton -- Appendix 1 Evolution of Isotopic Signatures during Fractionation Processes -- Problems -- 4. Radioactive and Radiogenic Isotopes -- 4.1. Radioactivity -- 4.2. The Radioactive Decay Law and its Applications -- 4.2.1. The Radioactive Decay Law -- 4.2.2. Disintegration without Simultaneous Production -- 4.2.3. Disintegration with Simultaneous Production -- 4.2.4. Definition of the Activity -- 4.3. The Long-Lived Radioactive Decay Systems -- 4.3.1. Strontium -- 4.3.2. Neodymium -- 4.3.3. Lead -- 4.3.4. Helium -- 4.4. The Uranium and Thorium Decay Chains -- 4.5. Cosmogenic Isotopes -- 4.5.1. The 14C Isotope -- 4.5.2. The 10Be Isotope -- 4.6. Artificial Isotopes -- Appendix 1 Integration of the Radioactivity Equation for a Closed System without Production Term -- Integration of the Radioactivity Equation for a Closed System with Production Term -- Calculation of the Mean Lifetime of an Isotope -- Problems -- 5. Box Models -- 5.1. One-Box Model -- 5.1.1. The Conservation Equation -- 5.1.2. Case of Enzyme Kinetics -- 5.1.3. Steady State -- 5.1.4. Residence Time -- 5.2. Dynamic Behavior of a Reservoir -- 5.2.1. Constant Forcing -- 5.2.2. Temporal Evolution of the Forcings -- 5.3. Box Models and Isotopic Tracers -- 5.3.1. Use of U and Th Decay Chains -- 5.3.2. Using the Isotopic Composition of a Tracer -- 5.3.3. Application Exercise: Ventilation of the Deep Waters in the Red Sea -- 5.4. Dynamics of Coupled Boxes -- 5.5. Mean Age, Residence Time and Reservoir Age of a Tracer -- Problems -- 6. Advection -- Diffusion Models --

6.1. An Infinitesimal Box -- 6.2. Advection -- 6.3. Molecular Diffusion -- 6.3.1. Random Walk -- 6.3.2. The Fick Law -- 6.3.3. Gas Diffusion at the Air-Sea Interface -- 6.4. Eddy Diffusion -- 6.5. The Full Conservation Equation -- 6.5.1. Example 1: Radium Transport in Coastal Waters -- 6.5.2. Example 2: Dispersion of SF_6 in the Thermocline -- 6.6. The Case of Sediment Transport -- Problems -- 7. Development and Limitations of Biological Activity in Surface Waters -- 7.1. Life Cycle in the Ocean -- 7.2. Development of the Biological Production in Surface Waters -- 7.3. Estimating the Primary Production -- 7.4. Global Distribution of Photosynthesis and Ocean Color -- 7.5. Iron Limitation -- 7.6. Silica Limitation -- 7.7. A CO_2 Limitation? -- 7.8. The Long-Term Limitation of the Production -- 7.9. Anthropogenic Impacts -- Problems -- 8. CO_2 Exchanges between the Ocean and the Atmosphere -- 8.1. The Global Carbon Cycle -- 8.2. The Partial Pressure of CO_2 in Seawater -- 8.2.1. Temperature Effect -- 8.2.2. Carbonate System Effect -- 8.2.3. Photosynthesis -- 8.2.4. Remineralization -- 8.2.5. The Formation of Calcium Carbonate ($CaCO_3$) -- 8.2.6. $CaCO_3$ Dissolution -- 8.2.7. Overall Effect on the Pumping of CO_2 -- 8.3. The Carbon Storage Capacity of the Ocean -- 8.4. Rate of CO_2 Transfer at the Air -- Sea Interface -- 8.5. Gas Equilibration Time between the Mixed Layer and the Atmosphere -- 8.5.1. Perturbation of Oxygen -- 8.5.2. Perturbation of the Carbonate System -- 8.5.3. Perturbation of the Isotopic Composition -- 8.6. Observation of the Anthropogenic Perturbation at the Ocean Surface -- 8.7. Global Estimate of Die Ocean -- Atmosphere Exchanges -- 8.8. Spread of the Anthropogenic Perturbation in the Deep Ocean -- Problems -- 9. The Little World of Marine Particles -- 9.1. Origin and Nature of Marine Particles -- 9.2. Marine

Particle Sampling -- 9.3. The Distribution of Particles -- 9.4. Particle Sinking -- 9.5. Changes of the Particle Flux with Depth -- 9.5.1. The Organic Matter Flux -- 9.5.2. The Mineral Phases -- 9.6. Estimation of the Particle Flux -- 9.6.1. 234Th and Irreversible "Scavenging" Models -- 9.6.2. Relations between Small and Large Particles -- 9.6.3. 230Th and Reversible Models -- 9.7. The Role of Margins -- 9.7.1. Boundary Scavenging -- 9.7.2. Boundary Exchange -- 9.8. The Distribution of Sediments on the Seafloor -- 9.9. The Diagenesis -- 9.10. Timescales and Sediment Fluxes -- Problems -- 10. Thermohaline Circulation -- 10.1. The Long Path of Deep Waters -- 10.2. The Rapid Progression of Transient Tracers -- 10.2.1. Deep Current Dynamics -- 10.2.2. Intensity of the Recirculation -- 10.3. 14C-Transient Tracer Comparison -- 10.4. The Contribution of 231Pa-230 Th -- 10.5. The Origin of the AABW -- 10.6. Closure of the Meridional Overturning Circulation -- Problems -- 11. Ocean History and Climate Evolution -- 11.1. The Origin of the Ocean -- 11.2. The First Traces of Life -- 11.3. The Rise of Oxygen -- 11.4. Geological Sequestration of CO2 -- 11.5. The Closure of the Panama Isthmus -- 11.6. The Last Glaciation -- 11.7. El Nino Exacerbated by Human Activity? -- 11.8. The Climate of the Future and the Ocean -- 11.9. The Expected Consequences -- Problems.

Subjects
Chemical oceanography.
Marine sediments.
Geochemistry.
Biogeochemistry.
Biogeochemistry.
Chemical oceanography.
Geochemistry.
Marine sediments.

Subject keywords Geochemistry, Biogeochemistry

	Geochemie, biogeochemie
Notes	Includes bibliographical references and index.

Marine pollution: sources, fate and effects of pollutants in coastal ecosystems

LCCN	2018934071
Type of material	Book
Personal name	Beiras, Ricardo, 1965- author.
Main title	Marine pollution: sources, fate and effects of pollutants in coastal ecosystems / Ricardo Beiras, University of Vigo.
Published/Produced	Amsterdam, Netherlands; Oxford, United Kingdom; Cambridge, MA: Elsevier; [2018]
Description	xxii, 385 pages: illustrations (some color); 24 cm
ISBN	9780128137369 (paperback) 0128137363 (paperback)
LC classification	GC1085 .B435 2018
Portion of title	Sources, fate and effects of pollutants in coastal ecosystems
Summary	"Marine Pollution: Sources, Fate and Effects of Pollutants in Coastal Ecosystems bring together the theoretical background on common and emerging marine pollutants and their effects on organisms (ecotoxicology). Written by a renowned expert in the field who is a researcher, teacher and advisor of national and international institutions on issues such as oil spills, water quality assessment and plastic pollution, this book offers a thorough account of the effects of pollutants on marine organisms, the relevant environmental regulations, and the public health implications, along with the biological tools advocated by the international institutions for marine pollution monitoring. Marine Pollution: Sources, Fate and Effects of Pollutants in Coastal Ecosystems presents information in a detailed and didactic manner, reviewing the latest scientific knowledge alongside examples of practical applications." -- Provided by publisher.

Contents	Part I. Pollutants in marine ecosystems. 1. Basic concepts; 2. Nonpersistent organic pollution; 3. Nonpersistent inorganic pollution; 4. Microbial pollution; 5. Liquid wastes: from self-purification to waste water treatment; 6. Plastics and other solid wastes; 7. Hydrocarbons and oil spills; 8. Persistent organic xenobiotics; 9. Trace metals and organometallic compounds -- Part II. Marine ecotoxicology. 10. Distribution of pollutants in marine environment; 11. Bioaccumulation; 12. Biotransformation; 13. Theory and practice of toxicology: toxicity testing; 14. Sublethal toxicity at the level of organism; 15. Effects of pollution on populations, communities, and ecosystems -- Part III. Monitoring and abatement of marine pollution. 16. Biological tools for monitoring: biomarkers and bioassays; 17. Marine pollution monitoring programs; 18. Pollution control: focus on emissions; 19. Pollution control: focus on Receiving waters
Subjects	Marine pollution. Pollutants. Marine ecology. Coastal ecology.
Notes	Includes bibliographical references and index.

Metal toxicity in higher plants

LCCN	2019054118
Type of material	Book
Main title	Metal toxicity in higher plants / [edited by Marco Landi].
Published/Produced	New York: Nova Science Publishers, Inc., [2019]
Description	viii, 257 pages: ill.; 25 cm.
ISBN	9781536167894 (hardcover) (adobe pdf)
LC classification	QK753.H4 M48 2019
Related names	Landi, Marco, editor.

Summary

"Metals are important environmental pollutants and their toxicity is a problem of increasing significance for ecological, nutritional, and environmental reasons. These pollutants, ultimately derived from a growing number of diverse anthropogenic sources (industrial effluents and wastes, urban runoff, sewage treatment plants, boating activities, agricultural fungicide runoff, domestic garbage dumps, and mining operations), have progressively affected more and more different ecosystems. Even agricultural lands are progressively becoming enriched of metals due to long-term use of phosphatic fertilizers, sewage sludge application, dust from smelters, industrial waste and bad watering practices in agricultural lands. Among these metals, Cu, Fe, Mn, Mo and Zn are pivotal micronutrients, while Ag, As, Cd, Cr, Hg, Pb, Sb and V and are non-essential for plants and have no known function as nutrients and seem to be more or less toxic to all plants and microorganisms. Sodium excess deserves particular attention. Several agricultural lands are indeed becoming familiar with the problem of salinization, due to the use of fresh water which contains a high level of NaCl or due to intensive fertilization, especially in arid and semi-arid environments characterized by poor rainfalls. Overall, the presence of both essential and non-essential metals in the atmosphere, soil and water, in excessive amounts, can cause serious problems to all organisms. Knowledge of plant-metal interactions is important for the safety of the environment, but also for reducing the risks associated with the introduction of trace metals into the food chain. Although intense research has been conducted during the last 30 years, many aspects remain to be clarified concerning the effect of metals in higher plants. Physiological and biochemical mechanisms adopted by plants to cope

	with metal excess and possible implications for phytoremediation of metal-contaminated soils are also discussed"-- Provided by publisher.
Subjects	Plants--Effect of heavy metals on. Metals--Toxicology.
Notes	Includes bibliographical references and index.
Additional formats	Online version: Metal toxicity in higher plants. New York: Nova Science Publishers, Inc., [2019] 9781536167900 (DLC) 2019054119
Series	Environmental science, engineering and technology, pollution science, technology and abatement

Metal toxicity in higher plants

LCCN	2019054119
Type of material	Book
Main title	Metal toxicity in higher plants / [edited by Marco Landi].
Published/Produced	New York: Nova Science Publishers, Inc., [2019]
Description	1 online resource
ISBN	9781536167900 (adobe pdf) (hardcover)
LC classification	QK753.H4
Related names	Landi, Marco, editor.
Summary	"Metals are important environmental pollutants and their toxicity is a problem of increasing significance for ecological, nutritional, and environmental reasons. These pollutants, ultimately derived from a growing number of diverse anthropogenic sources (industrial effluents and wastes, urban runoff, sewage treatment plants, boating activities, agricultural fungicide runoff, domestic garbage dumps, and mining operations), have progressively affected more and more different ecosystems. Even agricultural lands are progressively becoming enriched of metals due to long-term use of phosphatic fertilizers, sewage sludge application, dust from smelters, industrial waste and bad

	watering practices in agricultural lands. Among these metals, Cu, Fe, Mn, Mo and Zn are pivotal micronutrients, while Ag, As, Cd, Cr, Hg, Pb, Sb and V and are non-essential for plants and have no known function as nutrients and seem to be more or less toxic to all plants and microorganisms. Sodium excess deserves particular attention. Several agricultural lands are indeed becoming familiar with the problem of salinization, due to the use of fresh water which contains a high level of NaCl or due to intensive fertilization, especially in arid and semi-arid environments characterized by poor rainfalls. Overall, the presence of both essential and non-essential metals in the atmosphere, soil and water, in excessive amounts, can cause serious problems to all organisms. Knowledge of plant-metal interactions is important for the safety of the environment, but also for reducing the risks associated with the introduction of trace metals into the food chain. Although intense research has been conducted during the last 30 years, many aspects remain to be clarified concerning the effect of metals in higher plants. Physiological and biochemical mechanisms adopted by plants to cope with metal excess and possible implications for phytoremediation of metal-contaminated soils are also discussed"-- Provided by publisher.
Subjects	Plants--Effect of heavy metals on. Metals--Toxicology.
Notes	Includes bibliographical references and index. Description based on print version record and CIP data provided by publisher; resource not viewed.
Additional formats	Print version: Metal toxicity in higher plants New York: Nova Science Publishers, Inc., [2019] 9781536167894 (DLC) 2019054118
Series	Environmental science, engineering and technology, pollution science, technology and abatement

Metals in the brain: measurement and imaging

LCCN	2017937544
Type of material	Book
Main title	Metals in the brain: measurement and imaging / edited by Anthony R. White, Department of Pathology, University of Melbourne, Melbourne, VIC, Australia.
Published/Produced	New York, NY: Humana Press, [2017]
Description	xii, 270 pages: illustrations (some color); 26 cm.
ISBN	9781493969166 (alk. paper) 1493969161 (alk. paper)
LC classification	RC386.2 .M4835 2017
Related names	White, Anthony R., Dr., editor.
Summary	"Our understanding of how metals contribute to neural cell function and disease in the brain is rapidly evolving. A broad range of metals occur in the brain, from trace levels through to bulk amounts. Methods for measuring levels of metals and imaging metal localization are diverse and often prone to pitfalls associated with a lack of experience in the factors that affect metal detection. These include use of non-physiological metal treatments, fixation artifacts in tissues, mis-interpretation of data from metal-probes, and mis-understanding of the sensitivity and/or specificity of many techniques. This book is designed to bring together a well-considered selection of methods used by experienced metal researchers to aid investigators in their plans to examine metal levels or distribution in brain tissue or brain-derived cells and avoid many of the pitfalls that can lead to nonreproducible and artifact-prone data. The chapters following this introduction will provide a broad and where necessary in-depth coverage of the leading approaches used today for assessment of metal levels in brain, brain-derived cells, and in some cases serum (which can be used to provide important collorary information on neural metal

levels). Subsequent chapters provide a more focused examination of the methodology and issues associated with measuring some of the more commonly studied brain metals. It is hoped that whether researchers read all the chapters or just a selection targeting their metal or technique of interest, they can take away important new insights to provide accurate and reproducible metal measurement relevant to brain tissue. These improvements will help the field to establish and maintain metal regulation as a critical aspect of brain function in health and disease"--Introduction.

Contents Introduction and overview / Anthony R. White -- X-ray microscopy for detection of metals in the brain / Joanna F. Collingwood and Freddy Adams -- Imaging metals in the brain by laser ablation-inductively coupled plasma-mass spectrometry / Dominic J. Hare, Bence Paul, and Philip A. Doble -- Fluorescent probes for the analysis of labile metals in brain cells / Jacek L. Kolanowski, Clara Shen, and Elizabeth J. New -- Probing biological trace metals with fluorescent indicators / Christoph J. Fahrni, Daisy Bourassa, and Ryan Dikdan -- Microdissection of alzheimer brain tissue for the determination of focal manganese accumulation / Marcus W. Brazier -- Analysis of trace elements and metalloproteins in fractionated human brain samples using size exclusion inductively coupled mass spectrometry / Adam P. Gunn and Blaine R. Roberts -- Size fractionation of metal species from serum samples for studying element biodistribution in alzheimer's disease / Raúl González Domínguez -- Lead uptake and localization in glial cell cultures / Rola Barhoumi, Robert Taylor, and Evelyn Tiffany-Castiglioni -- Evaluating iron flux in the brain / Bruce X. Wong, Linh Q. Lam, Andrew Tsatsanis, and James A. Duce -- Current methods used to probe and quantify intracellular total and

	free Zn(II) dynamics, and subcellular distribution in cultured neurons / Yan Qin, Kyle R. Gee, Qiaoling Jin, Barry Lai, Cheng Qian, and Robert A. Colvin -- Monitoring intracellular Zn2+ using fluorescent sensors: facts and artifacts / Lech Kiedrowski -- Measuring changes in brain manganese or iron using magnetic resonance imaging (MRI) / Kimberly L. Desmond and Nicholas A. Bock -- Treatment and measurement of metals in brain cell cultures / Xin Yi Choo and Alexandra Grubman.
Subjects	Brain--Metabolism. Brain--Effect of metals on. Neurotoxicology. Brain chemistry. Brain--metabolism. Metals--analysis. Diagnostic Imaging--methods. Nervous System Diseases--chemically induced.
Form/Genre	Laboratory Manuals.
Notes	Includes bibliographical references and index.
Series	Neuromethods, 0893-2336; 124 Springer protocols Neuromethods; 124. 0893-2336 Springer protocols (Series) 1949-2448

Phytoremediation: methods, management and assessment

LCCN	2018003180
Type of material	Book
Uniform title	Phytoremediation (Nova Science Publishers)
Main title	Phytoremediation: methods, management and assessment / Vladimir Matichenkov, editor.
Published/Produced	New York: Nova Science Publishers, [2017]
Description	xxxv, 340 pages; 26 cm.
ISBN	9781536131161 (hardcover) 1536131164 (ebook)
LC classification	TD192.75 .P475 2017
Related names	Matichenkov, Vladimir, editor.

Summary

Phytoremediation is one of the cost effective procedures for cleaning or removing pollutants from soil or water matrices. Over the last few decades, thousands of publications about different aspects of phytoremediation were created. This massive amount of data requires systematization, classification and ordering. On the other hand, many aspects of phytoremediation are not elaborated properly. This current book contains classical and specific literature reviews, new approaches in phytoremediation techniques and new areas for the realization of phytoremediation, which is not related with traditional pollutants as heavy metals. The first chapter is a classical review about using high biomass non-hyperaccumulator plant species for remediation of the toxic metal polluted soils. This chapter provides a lot of information to help readers better understand physiological, biochemical, and molecular aspects of such regular plant species used to increase phytoremediation efficiency. The second chapter ("Arsenic, the Silent Threat: New Phytoremediation Strategies for Contaminated Soils and Waters") provides information about one of the most dangerous toxic metalloids, which is known as arsenic. This chapter aims to describe the current state of research and advances of knowledge concerning the phytoremediation of arsenic-polluted areas, focusing on mechanisms of As response in plants and the different strategies to cope with the metalloid, tolerance mechanisms that can be used to modify arsenic uptake, transport or detoxification in order to improve phytoremediation efficiency. The selection of the right and most effective plant for phytoremediation is one of the most important parts of this book-- Provided by publisher.

Contents

Toxic Metal Phytoremediation Using High Biomass Non-Hyperaccumulator Crops: New Possibilities for Bioenergy Resources / Lucas Anjos Souza, Liliane Santos Camargos and Márcia Eugenia Amaral Carvalho, Campus Rio Verde, Instituto Federal de Educação, São Paulo, Brazil, and others -- Arsenic, the Silent Threat: New Phytoremediation Strategies for Contaminated Soils and Waters / Cristina Navarro, Sarra Arbaoui, Cristian Mateo, Taoufik Bettaieb and Antonio Leyva, Department of Plant Molecular Genetics, Centro Nacional de Biotecnología, Consejo Superior de Investigaciones Científicas (CSIC), Madrid, Spain, and others -- The Use of Forage Grasses for the Phytoremediation of Heavy Metals: Plant Tolerance Mechanisms, Classifications, and New Prospects / Flávio H.S. Rabêlo, Lucélia Borgo and José Lavres, University of São Paulo, Center for Nuclear Energy in Agriculture, Division of Agroindustrial Productivity and Food, Piracicaba, São Paulo, Brazil, and others -- Halophytes as Valuable Tools in the Phytoremediation of Coastal Environments Contaminated by Trace Metals / Carmen A. Pedro, Márcia S.S. Santos, Susana M.F. Ferreira, and Sílvia C. Gonçalves, Marine and Environmental Sciences Centre, ESTM, Polytechnic Institute of Leiria, Peniche, Portugal -- Chelate-Enhanced Phytoremediation of Pb and Cu in Single and Co-Contaminated Soils Using Tagetes Erecta / Chibuike Chigbo and Oyinkepreye Koufa, Department of Environmental Technology, Federal University of Technology, Owerri, Nigeria, and others -- The Management of Heavy Metal Uptake and Transport by Plants under Si Application / Yuqiao Liu, Xiao Wei, Dandan Zhao, Pengbo Zhang, Qiang Zhan, Bo Xu, Elena Bocharnikova, and Vladimir Matichenkov, Hunan Economic Geography Institute, Changsha, China, and others -

- The Role of Vascular Plants in the Phytoremediation of Fly Ash Deposits / Gordana M. Gajic and Pavle. Pavlovic, Department of Ecology, Institute for Biological Reasearch Sinia Stankovic, University of Belgrade, Belgrade, Serbia -- Phytoremediation Using Microalgae: Techniques and Perspectives / Alexis Hernández-Pérez, Humberto Mattos, Juan L. Ramos-Suárez, Desarrollo de Proyectos e Investigación Científica, Cokaru I+D, Santiago, Chile, and others -- Salinity Effects on Plants and Their Desalination Potential in Aquatic Environments / Maria. A. de C. Gomes and Rachel A. Hauser-Davis, Laboratório de Ciências Ambientais, Centro de Biociências e Biotecnologia, Universidade Estadual do Norte Fluminense Darcy Ribeiro, Campos dos Goytacazes, RJ, Brazil, and others.

Subjects Phytoremediation.

Notes Includes bibliographical references and index.

Additional formats Online version: Hauppauge, New York: Nova Science Publishers, Inc., [2017] 9781536131178 (DLC) 2018014894

Series Environmental research advances

Recent advances in Applied Microbiology

LCCN 2019764232

Type of material Book

Main title Recent advances in Applied Microbiology / edited by Pratyoosh Shukla.

Edition 1st ed. 2017.

Published/Produced Singapore: Springer Singapore: Imprint: Springer, 2017.

Description 1 online resource (XIII, 290 pages 129 illustrations, 56 illustrations in color.)
text file PDF

ISBN 9789811052750

Related names Shukla, Pratyoosh, editor.

Summary

This book is a one-stop reference resource, presenting recent research in various emerging areas of microbiology, including microbial biotechnology, microbes in health, microbial interactions, agricultural microbiology and computational approaches. Recent discoveries in microbiology have created a great deal of interest among researchers around the globe, and as as such the book discusses a number of important research topics, such as microbial enzymes and nanoparticles, bacterial polyhydroxyalkanoates, biosurfactant aided bioprocessing, autophagy and microbial pathogenesis, multidrug resistant bacteria, probiotics, rhizosphere, metal tolerant bacteria, plant- beneficial environmental bacteria and therapeutic applications of fungal chondroitinase. It serves as a valuble resource for masters, doctoral and postdoctoral researchers in life sciences, as well as scientists involved in various interdisciplinary research areas. It also provides useful material for higher-level graduate courses in microbiology and biotechnology.

Contents

Prevalence of Multidrug Resistant Bacteria in River Cauvery and Computational Virtual Screening for Natural Inhibitors against MDR Genes -- 9. Probiotics for human health: Current Progress and Applications -- Section 3: Microbial Interactions.- 10. Functional characterization of phosphate solubilizing bacteria from coastal rice soils of Odisha -- 11. Tolerance of filamentous fungi and bacteria in soil contaminated with trace metals -- 12. Transformation, purification and quantification of soy isoflavone from Lactobacillus sp., and Bifidobacterium sp. -- 13. Processing of pearl millet to prepare ready to eat (rte) extruded healthy snacks -- 14. Studies on screening, isolation and optimization of fungal chondroitinase -- 15. Citricoccus zhacaiensis - A novel osmotolerant plant growth promoting actinobacterium -- Section 4: Computational approaches in microbiology -- 16. In-silico design and antibody response to peptide sequences from protective antigen and lethal factor toxins of Bacillus anthracis -- 17. Microbial enzyme engineering: applications and perspectives.

Subjects Bacteriology.
Biomedical engineering.
Medical microbiology.
Microbiology.
Biomedical Engineering/Biotechnology.
Applied Microbiology.
Bacteriology.
Food Microbiology.
Medical Microbiology.

Notes Description based on publisher-supplied MARC data.

Additional formats Print version: Recent advances in applied microbiology 9789811052743 (DLC) 2017954324
Printed edition: 9789811052743
Printed edition: 9789811052767
Printed edition: 9789811353567

The Rivers of Greece: Evolution, Current Status and Perspectives

LCCN	2019755090
Type of material	Book
Main title	The Rivers of Greece: Evolution, Current Status and Perspectives / edited by Nikos Skoulikidis, Elias Dimitriou, Ioannis Karaouzas.
Edition	1st ed. 2018.
Published/Produced	Berlin, Heidelberg: Springer Berlin Heidelberg: Imprint: Springer, 2018.
Description	1 online resource (XIV, 439 pages 110 illustrations, 90 illustrations in color.) text file PDF
ISBN	9783662553695
Related names	Dimitriou, Elias. editor. Karaouzas, Ioannis. editor. Skoulikidis, Nikos. editor.
Summary	This volume provides essential information on the origin and evolution of Greek rivers, as well as their ecological and anthropogenic characteristics. The topics covered include geomythology, biogeography, hydrology, hydrobiology, hydrogeochemistry, geological and biogeochemical processes, anthropogenic pressures and ecological impacts, water management - both in the antiquity and today - and river restoration. The book is divided into four parts, the first of which explores the importance of rivers for ancient Greek civilization and the natural processes affecting their evolution during the Holocene. In the second part, the hydrological, hydrochemical and biological features of Greek rivers and the unique biogeographical characteristics that form the basis for their high biodiversity and endemism are highlighted, while the third part comprehensively discusses the impacts of environmental pollution on the structure and function of Greek river ecosystems. In turn, the final part describes the current socio-economic factors in Greece that are

affecting established water management practices, the application of ecohydrological approaches in restoring fragmented rivers, and the lessons learned from restoring aquatic ecosystems in general as a paradigm for understanding and minimizing anthropogenic impacts on water resources, at the Mediterranean scale. Given the breadth and depth of its coverage, the book offers an invaluable source of information for researchers, students and environmental managers alike.

Contents Ancient Greece and Water: Climatic changes, extreme events, water management and rivers in Ancient Greece -- Natural Processes versus Human Impacts during the Holocene - A case study of Aliakmon River Delta -- The biogeographic characteristics of the river basins of Greece -- The state and origin of river water composition in Greece -- Long-Term Hydrologic Trends in the Main Greek Rivers: A Statistical Approach -- Agro-Industrial Wastewater Pollution in Greek River Ecosystems -- Overview of the pesticide residues in Greek Rivers: Occurrence and Environmental Risk Assessment -- Geochemical processes of trace metals in fresh- saline water interfaces. The cases of Louros and Acheloos Estuaries -- The Evrotas River Basin - 10 years of river monitoring -- Review on macroinvertebrate assemblages and biological status of rivers in Northern and Central Greece -- Socio-Economics and Water Management: Revisiting the Contribution of Economics in the Implementation of The Water Framework Directive in Greece -- Environmental Impacts of Large-scale Hydropower Projects and Applied Ecohydrology Solutions for Watershed Restoration: The case of Nestos River, Northern Greece -- Restoration of rivers and wetlands in Greece: Insights Lessons from biodiversity conservation initiatives.

Subjects Freshwater.

	Geochemistry. Marine sciences. Water pollution. Water quality. Marine & Freshwater Sciences. Geochemistry. Water Quality/Water Pollution.
Notes	Description based on publisher-supplied MARC data.
Additional formats	Print version: The rivers of Greece: evolution, current status and perspectives 9783662553671 (DLC) 2017954950 Printed edition: 9783662553671 Printed edition: 9783662553688 Printed edition: 9783662572313
Series	The Handbook of Environmental Chemistry, 1867-979X; 59 The Handbook of Environmental Chemistry, 1867-979X; 59

Trace Metal Biogeochemistry and Ecology of Deep-Sea Hydrothermal Vent Systems

LCCN	2019750431
Type of material	Book
Main title	Trace Metal Biogeochemistry and Ecology of Deep-Sea Hydrothermal Vent Systems / edited by Liudmila L. Demina, Sergey V. Galkin.
Edition	1st ed. 2016.
Published/Produced	Cham: Springer International Publishing: Imprint: Springer, 2016.
Description	1 online resource (XII, 210 pages) text file PDF
ISBN	9783319413402
Related names	Demina, Liudmila L. editor. Galkin, Sergey V. editor.
Summary	This volume synthesizes the relevant data that is fundamental to our understanding of trace metal biogeochemistry and the ecology of biological

communities of deep-sea vent systems. It presents the combined results of biological and geochemical research and analyzes the microdistribution of animals and the spatial structure of vent communities. Careful consideration is given to the export of iron and other trace metals from hydrothermal vents. The environmental conditions to be found in deep-sea hydrothermal community habitats, along with the trace metal behavior in biotope water are characterized and the sources and forms of trace metals taken up by dominant hydrothermal vent animals are discussed. Special attention is paid to the poorly investigated deep biosphere of the sub-seafloor igneous crust. The book is illustrated with a wealth of exceptional deep-sea photos taken by the manned submersible "Mir", and a dedicated chapter focuses on the role of deep manned submersibles in ocean research. The book will be of interest to researchers and students in the fields of oceanography, geochemistry, biology, the environmental sciences and marine ecology.

Contents Introduction -- The export of iron and other trace metals from hydrothermal vents and the impact on their marine biogeochemical cycle -- Geologic-geochemical and ecological characteristics of selected hydrothermal vent areas -- Trace metals in water of deep-sea hydrothermal biotopes -- Structure of hydrothermal vent communities -- Sources and forms of trace metals taken up by hydrothermal vent mussels, possible adaptation and mediation strategies -- Factors controlling the trace metal distribution in hydrothermal vent organisms -- The deep biosphere of the subseafloor igneous crust -- Manned submersibles Mir and the worldwide research of hydrothermal vents -- Conclusions.

Subjects Analytical chemistry.

	Aquatic ecology. Environmental chemistry. Geobiology. Geochemistry. Environmental Chemistry. Analytical Chemistry. Biogeosciences. Freshwater & Marine Ecology. Geochemistry.
Notes	Description based on publisher-supplied MARC data.
Additional formats	Print version: Trace metal biogeochemistry and ecology of deep-sea hydrothermal vent systems 9783319413389 (DLC) 2016946030 Printed edition: 9783319413389 Printed edition: 9783319413396 Printed edition: 9783319823287
Series	The Handbook of Environmental Chemistry, 1867-979X; 50 The Handbook of Environmental Chemistry, 1867-979X; 50

Trace metals: evolution, environmental and ecological significance

LCCN	2017040627
Type of material	Book
Main title	Trace metals: evolution, environmental and ecological significance / Mildred McCarthy, editor.
Published/Produced	New York: Nova Science Publishers, [2017]
Description	1 online resource.
ISBN	9781536124101 (ebook)
LC classification	TD196.M4
Related names	McCarthy, Mildred, editor.
Subjects	Metals--Environmental aspects. Trace elements--Environmental aspects. Trace elements in water. Soils--Trace element content.
Notes	Includes bibliographical references and index.

	Description based on print version record and CIP data provided by publisher.
Additional formats	Print version: Trace metals Hauppauge, New York: Nova Science Publishers, Inc., [2017] 9781536124033 (DLC) 2017036257
Series	Environmental health - physical, chemical and biological factors

Trace metals: evolution, environmental and ecological significance

LCCN	2017036257
Type of material	Book
Main title	Trace metals: evolution, environmental and ecological significance / editor, Mildred McCarthy.
Published/Produced	New York: Nova Science Publishers, [2017]
Description	xii, 136 pages; 23 cm.
ISBN	9781536124033 (softcover)
LC classification	TD196.M4 T72 2017
Related names	McCarthy, Mildred, editor.
Subjects	Metals--Environmental aspects. Trace elements--Environmental aspects. Trace elements in water. Soils--Trace element content.
Notes	Includes bibliographical references and index.
Additional formats	Online version: Trace metals Hauppauge, New York: Nova Science Publishers, Inc., [2017] 9781536124101 (DLC) 2017040627
Series	Environmental health - physical, chemical and biological factors

Trace metals and infectious diseases

LCCN	2014039756
Type of material	Book
Main title	Trace metals and infectious diseases / edited by Jerome O. Nriagu and Eric P. Skaar.
Published/Produced	Cambridge, Massachusetts: The MIT Press, [2015]
Description	xii, 488 pages: illustrations (some color); 24 cm.
ISBN	9780262029193 (hardcover: alk. paper)
LC classification	TD196.T7 T725 2015

Related names Nriagu, Jerome O., editor.
Skaar, Eric P. (Eric Patrick), editor.
Subjects Trace elements--Toxicology.
Trace elements--Environmental aspects.
Communicable diseases.
Notes Includes bibliographical references and index.
Series Strungmann Forum reports

Trace metals in a tropical Mangrove wetland: chemical speciation, ecotoxicological relevance and remedial measures

LCCN 2017944178
Type of material Book
Personal name Sarkar, Santosh Kumar.
Main title Trace metals in a tropical Mangrove wetland: chemical speciation, ecotoxicological relevance and remedial measures / Santosh Kumar Sarkar.
Published/Produced New York, NY: Springer Berlin Heidelberg, 2017.
ISBN 9789811027925

Trace Metals in a Tropical Mangrove Wetland: Chemical Speciation, Ecotoxicological Relevance and Remedial Measures

LCCN 2019761213
Type of material Book
Personal name Sarkar, Santosh Kumar, author.
Main title Trace Metals in a Tropical Mangrove Wetland: Chemical Speciation, Ecotoxicological Relevance and Remedial Measures / by Santosh Kumar Sarkar.
Edition 1st ed. 2018.
Published/Produced Singapore: Springer Singapore: Imprint: Springer, 2018.
Description 1 online resource (XIX, 247 pages 37 illustrations, 26 illustrations in color.)
text file PDF
ISBN 9789811027932
Summary This book offers a comprehensive and accessible guide covering various aspects of trace metal contamination in abiotic and biotic matrices of an iconic Indian tropical mangrove wetland -

Sundarban. Divided into nine chapters, the book begins by discussing the fundamental concepts of sources, accumulation rate and significance of trace metal speciation, along with the impact of multiple stressors on trace metal accumulation, taking into account both tourist activities and the exacerbating role of climate change. The second chapter presents a detailed account of the sampling strategy and preservation of research samples, followed by exhaustive information on sediment quality assessment and ecological risk, instrumental techniques in environmental chemical analyses, quality assurance and quality control, along with the Sediment Quality Guidelines (SQGs). Using raw data, the sediment quality assessment indices (e.g., pollution load index, index of geoaccumulation, Nemerow Pollution Load Index et cetera) and conventional statistical analyses are worked out and interpreted precisely, allowing students to readily evaluate and interpret them. This is followed by chapters devoted to trace metal accumulation in sediments and benthic organisms, as well as acid-leachable and geochemical fractionation of trace metals in sediments. The book then focuses on chemical speciation of butylin and arsenic in sediments as well as macrozoobenthos (polychaetous annelids). Finally, potential positive role of the dominant mangrove Avicennia in sequestering trace metals from rhizosediments of Sundarban Wetland is elaborately discussed. This timely reference book provides a versatile and in-depth account for understanding the emerging problems of trace metal contamination - issues that are relevant for many countries around the globe.

Contents 1. Introduction -- 2. Materials and Methods -- 3. Trace Element Contamination in Surface Sediment of Sundarban Wetland -- 4. Total and Acid-leachable Trace Metals in Surface Sediments along

	Hooghly (Gangels) River Estuary, India -- 5. Bioaccumulation of Trace Metals in Macrozoobenthos of Sundarban Wetland -- 6. Geochemical Speciation and Risk Assessment of Trace Metals in Sediments of Sundarban Wetland -- 7. Organotin Compounds in Surficial Sediments of Sundarban Wetland and Adjacent Coastal Regions -- 8. Arsenic Speciation in Sediments and Representative Biota of Sundarban Wetland -- 9. Phytoremediation of Trace Metals by Mangrove Plants in Sundarban Wetland.
Subjects	Climate change. Ecotoxicology. Environmental chemistry. Environmental management. Freshwater. Marine sciences. Water pollution. Marine & Freshwater Sciences. Climate Change/Climate Change Impacts. Ecotoxicology. Environmental Chemistry. Environmental Management. Waste Water Technology / Water Pollution Control / Water Management / Aquatic Pollution.
Notes	Description based on publisher-supplied MARC data.
Additional formats	Print version: Trace metals in a tropical Mangrove wetland: chemical speciation, ecotoxicological relevance and remedial measures 9789811027925 (DLC) 2017944178 Printed edition: 9789811027925 Printed edition: 9789811027949 Printed edition: 9789811097065

Trace metals in the environment and living organisms: the British isles as a case study

LCCN	2018017273

Type of material Book
Personal name Rainbow, P. S., author.
Main title Trace metals in the environment and living organisms: the British isles as a case study / Philip S. Rainbow, Natural History Museum.
Published/Produced Cambridge; New York, NY: Cambridge University Press, 2018.
Description xv, 742 pages: illustrations (some color), maps; 26 cm
ISBN 9781108470933 (hardback)
9781108456869 (paperback)
LC classification QH545.M45 R35 2018
Subjects Metals--Environmental aspects--British isles.
Notes Includes bibliographical references (pages 671-719) and index.

Treatment Wetlands for Environmental Pollution Control
LCCN 2019738479
Type of material Book
Personal name Obarska-Pempkowiak, Hanna. author.
Main title Treatment Wetlands for Environmental Pollution Control / by Hanna Obarska-Pempkowiak, Magdalena Gajewska, Ewa Wojciechowska, Janusz Pempkowiak.
Edition 1st ed. 2015.
Published/Produced Cham: Springer International Publishing: Imprint: Springer, 2015.
Description 1 online resource (XIII, 169 pages 60 illustrations, 8 illustrations in color.)
text file PDF
ISBN 9783319137940
Related names Gajewska, Magdalena author.
Wojciechowska, Ewa author.
Pempkowiak, Janusz. author.
Summary The aim of this book is to present an overview of the state of the art with regard to the function, application and design of TWSs in order to better protect surface water from contamination.

Accordingly, it also presents applications of constructed wetlands with regard to climatic and cultural aspects. The use of artificial and natural treatment wetland systems (TWSs) for wastewater treatment is an approach that has been developed over the last thirty years. Europe is currently home to roughly 10,000 constructed wetland treatment systems (CWTSs), which simulate the aquatic habitat conditions of natural marsh ecosystems; roughly 3,500 systems are in operation in Germany alone. TWSs can also be found in many other European countries, for example 200 - 400 in Denmark, 400 - 600 in Great Britain, and circa 1,000 in Poland. Most of the existing systems serve as local or individual household treatment systems. CWTSs are easy to operate and do not require specialized maintenance; further, no biological sewage sludge is formed during treatment processes. As TWSs are resistant to fluctuations in hydraulic loads, they are primarily used in rural areas as well as in urbanized areas with dispersed habitats, where conventional sewer systems and central conventional wastewater treatment plants (WWTPs) cannot be applied due to the high costs they would entail. TWSs are usually applied at the 2nd stage of domestic wastewater treatment, after mechanical treatment, and/or at the 3rd stage of treatment in order to ensure purification of effluent from conventional biological reactors and re-naturalization. New applications of TWSs include rainwater treatment as well as industrial and landfill leachate treatment. TWSs are well suited to these fields, as they can potentially remove not only organic matter and nitrogen compounds but also trace metals and traces of persistent organic pollutants and pathogens. Based on the practical experience gathered to date, and on new research regarding the processes and mechanisms of

	pollutant removal and advances in the systems properties and design, TWSs continue to evolve.
Contents	Introduction -- Characteristic of hydrophytes method -- Type of treatment wetlands -- Domestic wastewater treatment -- The quality of the outflow from conventional WWTPs and treatment wetlands systems -- Storm water treatment in TWs -- Landfill leachate treatment -- Reject water from digested sludge centrifugation in HTWs -- Sewage sludge stabilization.
Subjects	Hydrogeology. Water quality. Water pollution. Hydrogeology. Water Quality/Water Pollution. Waste Water Technology / Water Pollution Control / Water Management / Aquatic Pollution.
Notes	Description based on publisher-supplied MARC data.
Additional formats	Printed edition: 9783319137957 Printed edition: 9783319137933 Printed edition: 9783319385549
Series	GeoPlanet: Earth and Planetary Sciences, 2190-5193 GeoPlanet: Earth and Planetary Sciences, 2190-5193

Wetland Science: Perspectives From South Asia

LCCN	2019767264
Type of material	Book
Main title	Wetland Science: Perspectives From South Asia / edited by B. Anjan Kumar Prusty, Rachna Chandra, P. A. Azeez.
Edition	1st ed. 2017.
Published/Produced	New Delhi: Springer India: Imprint: Springer, 2017.
Description	1 online resource (XXVI, 587 pages 109 illustrations, 60 illustrations in color.) text file PDF

ISBN	9788132237150
Related names	Azeez, P. A, editor. Chandra, Rachna, editor. Prusty, B. Anjan Kumar, editor.
Summary	This book is an attempt to acknowledge the discipline 'wetland science' and to consolidate research findings, reviews and synthesis articles on different aspects of the wetlands in South Asia. The book presents 30 chapters by an international mix of experts in the field, who highlight and discuss diverse issues concerning wetlands in South Asia as case studies. The chapters are divided into different themes that represent broad issues of concern in a systematic manner keeping in mind students, researchers and general readers at large. The book introduces readers to the basics and theory of wetland science, supplemented by case studies and examples from the region. It also offers a valuable resource for graduate students and researchers in allied fields such as environmental studies, limnology, wildlife biology, aquatic biology, marine biology, and landscape ecology. To date the interdisciplinary field 'wetland science' is still rarely treated as a distinct discipline in its own right. Further, courses on wetland science aren't taught at any of the world's most prestigious universities; instead, the topics falling under this discipline are generally handled under the disciplines 'ecology' or under the extremely broad heading of 'environmental studies'. It is high time that 'Wetland Science' be acknowledged as an interdisciplinary sub-discipline, which calls for an attempt to consolidate its various subtopics and present them comprehensively. Thus, this book also serves as a reference base on wetlands and facilitates further discussions on specific issues involved in safeguarding a sustainable future for the wetland habitats of this region.

Contents 1. Wetlands Introductory -- i. Wetlands: origin and typology -- ii. Wetlands: forms and distribution -- 2. Bio-geochemical issues, Limnology and Hydrology -- i. Water chemistry -- ii. Sediment quality -- Biomass and Productivity -- iii. Decomposition and Mineralization -- iv. Water discharge and sediment loading -- volume Wetlands and Groundwater -- vi. Water Budgeting -- 3. Wetland biodiversity -- i. Biodiversity in Indian inland wetlands -- ii. Biodiversity in Indian coastal wetlands -- iii. Intertidal fauna and their role in coastal wetlands -- iv. Role of mangroves in coastal wetland protection -- 4. Current issues and Climate change -- i. Urbanization and wetlands -- ii. Industrialization and wetlands -- iii. Constructed wetlands -- iv. Carrying capacity volumeCarbon stock and sequestration potential of wetlands in India -- vi. Methane and other GHG emissions from wetlands -- 5. Ecosystem Goods and Services -- i. People's dependency on wetlands -- ii. Role of wetlands in economy of adjoining areas -- iii. Agriculture, fisheries, animal husbandry -- iv. Services and valuation of wetlands -- 6. Mapping of wetlands with respect to: -- i. Wildlife -- ii. Nutrient availability and transport -- iii. Carbon stock -- iv. Resource use and dependency: Socio-economics -- volume Ecosystem goods and services -- 7. Modeling and simulation -- i. Modeling wetland systems and processes -- ii. Mass balance model of chemical fate in wetlands -- iii. Modeling of chemical speciation of trace metals in wetlands -- 8. Wetland restoration -- i. Tools and techniques of wetland restoration -- ii. Species re-introduction and restoration -- iii. Seed-bank: implications for restoration -- 9. Wetlands in India: Legislation and Policy framework -- i. Legal aspects and provisions for wetland management -- ii. Policy framework and issues concerning wetlands -- iii. Institutional

	framework for wetland conservation and management -- 10. Wetland Science in India: Gaps and Futuristic -- i. Wetland Science: The Indian Experience -- ii. Research Matrix for Wetland studies in India -- iii. Gaps and Futuristic.
Subjects	Aquatic ecology . Environmental economics. Environmental law. Environmental policy. Environmental Law/Policy/Ecojustice. Environmental Economics. Freshwater & Marine Ecology.
Notes	Description based on publisher-supplied MARC data.
Additional formats	Print version: Wetland science: perspectives from South Asia 9788132237136 (DLC) 2017933841 Printed edition: 9788132237136 Printed edition: 9788132237143 Printed edition: 9788132238935

Index